Biosemiotics

Volume 15

Aims and Scope of the Series

Combining research approaches from biology, semiotics, philosophy and linguistics, the field of biosemiotics studies semiotic processes as they occur in and among living systems. This has important implications and applications for issues ranging from natural selection to animal behaviour and human psychology, leaving biosemiotics at the cutting edge of the research on the fundamentals of life.

The Springer book series *Biosemiotics* draws together contributions from leading scholars in international biosemiotics, producing an unparalleled series that will appeal to all those interested in the origins and evolution of life, including molecular and evolutionary biologists, ecologists, anthropologists, psychologists, philosophers and historians of science, linguists, semioticians and researchers in artificial life, information theory and communication technology.

More information about this series at http://www.springer.com/series/7710

Paul Cobley

Cultural Implications of Biosemiotics

Springer

Paul Cobley
Distinguished Visiting Professor,
School of Foreign Languages
and Cultures
Nanjing Normal University
Nanjing, China

President of the International Association
for Semiotic Studies; and Professor in
Language and Media, School of Media
and Performing Arts
Middlesex University
London, UK

ISSN 1875-4651 ISSN 1875-466X (electronic)
Biosemiotics
ISBN 978-94-024-0857-7 ISBN 978-94-024-0858-4 (eBook)
DOI 10.1007/978-94-024-0858-4

Library of Congress Control Number: 2016949398

Printed on acid-free paper

This Springer imprint is published by Springer Nature
The registered company is Springer Science+Business Media B.V. Dordrecht

For
John Deely

Acknowledgments

It is impossible to work for any extended period of time in any field unless you have the support of your colleagues. Such support is given for free and often goes unacknowledged. I take this opportunity to say heartfelt thanks and salute the following colleagues who made this book possible:

Myrdene Anderson, Richard Appignanesi, Chris Arning, Baranna Baker, Kristian Bankov, Marcello Barbieri, Merja Bauters, (the late and sadly missed) Jeff Bernard, Karin Boklund-Lagopoulou, Priscila Borges, Søren Brier, Sara Cannizzaro, James Carney, Vincent Colapietro, Marcel Danesi, (the volume's dedicatee, who originally encouraged me to write this book) John Deely, Jonathan Delafield-Butt, Per Durst-Andersen, Malcolm Evans, Xiaoli Fang, Don Favareau, José Enrique Finol, Alex Gordon, Anne Hénault, Jesper Hoffmeyer, Karel Kleisner, Evangelos Kourdis, Kalevi Kull, Alexandros Lagopoulos, Yunhee Lee, Christina Ljungberg, David Machin, Kobus Marais, Timo Maran, (the late and remarkable) Solomon Marcus, Anton Markoš, Dario Martinelli, Floyd Merrell, Danuta Mirka, Ivan Mladenov, Lucia Neva, Grigoris Paschalidis, Pengjia, Susan Petrilli, Augusto Ponzio, Anti Randviir, Alexei Sharov, Roger Sabin, Peter J. Schulz, (the late and irreplaceable) Tom Sebeok, Farouk Seif, Gary Shank, Alexei Sharov, Frederik Stjernfelt, Kim Sung-do, Eero Tarasti, Daina Teters, Torkild Thellefsen, Morten Tønnessen, Peeter Torop, Barendt van Heusden, Stéphanie Walsh Matthews, Gloria Withalm, Evripides Zantides and Yiheng Zhao.

Special thanks go to my colleagues/friends at Space Doctors: particularly, Gareth Lewis, Fiona McNae, Pavla Pasekova, Stephen Seth;

To colleagues/friends at Nanjing Normal University: especially, Hongbing Yu, Jie Zhang, Yongxiang Wang, Zhijun Yan;

To colleagues/friends at Middlesex University: particularly those with whom I work most closely, my inspirational and supportive Dean, Carole-Anne Upton, and my congenial partner in research administration, Vida Midgelow;

At Springer: Catherine Cotton, then Ineke Ravesloot, Iza Witkowska and Nel van der Werf – thanks, above all, for your patience. Plus the series editors whose work I value highly.

Needless to say, I'm grateful to all of my colleagues/friends in semiotics and biosemiotics. Sorry if I have forgotten anyone who should be included in this list.

Still, special mention must go to Terry Deacon who, at a crucial stage, provided me with material which was instrumental in the book coming to fruition.

And, socially, Adam Briggs who provided me with victuals that helped maintain at least a modicum of sanity.

But the biggest thank you of all goes to Elsie, Stan and Alison who put up with a major disruption to the work/life balance.

Small parts of this volume appeared in somewhat different form in other publications.

I would like to thank the following for permission to reproduce extracts from the following articles in this book: Lawrence and Wishart for 'Semioethics, voluntarism and antihumanism' *New Formations* 62: 44–60 (Autumn 2007); de Gruyter for extracts from 'Codes and coding: Sebeok's zoosemiotics and the dismantling of the fixed-code fallacy' *Semiotica* 198: 32–45 (2014); Springer for extracts from 'The cultural implications of biosemiotics' *Biosemiotics* 3 (2): 225–244 (2010); the Semiotic Society of America for extracts from 'Enhancing survival by not enhancing survival: Sebeok's semiotics and the ultimate paradox of modelling", SSA Sebeok Fellow Address 3 October Seattle, *The American Journal of Semiotics* 30 (3–4): 191–204 (2014) and 'What the humanities are for – a semiotic perspective', *The American Journal of Semiotics* 30 (3–4): 205–228 (2014); and MDPI for extracts from 'Cybersemiotics and human modelling' *Entropy* 12: 2045–2066 (2010).

Contents

Introduction

This book is aimed at a number of different targets. The reason it exists is because I and a number of people with whom I have discussed the matter, think that the implications of biosemiotics are important but have not reached some of the people who need to hear about them. So, firstly, the volume is aimed at an audience outside biosemiotics and semiotics, in the humanities and social sciences principally, who might welcome some comments on the possible benefits to their subject area from a relatively new field. It is possible that they misconstrue that field as being concerned only with matters of relevance to certain parts of the sciences and/or that the field is none of their concern. Possibly, they have an impression of biosemiotics as something esoteric. Hopefully, they will find some of the argumentation in biosemiotics to be convincing for their own intellectual concerns. In my own discipline – communications, comprising media, language and cultural studies – theory such as that represented by biosemiotics has not had a prosperous time in the academy over the last couple of decades. The publishing bonanza attendant on 'postmodernism' nearly 30 years ago seems to have represented a last gasp of 'grand theory' in the arts, humanities and social sciences. So, if this volume is read by anyone who aligns themselves with those latter fields of interest, I will be happy. If they see this book as supplementing their interest in theory, contributing a couple of biosemiotic ideas to their existing repertoire, leading them to a few interesting references, or – the philosopher's stone – they become converts to biosemiotics: I'll be extremely happy.

In addition to the more modest aims, here, the volume is also targeted at fellow workers in biosemiotics. It is hoped that it will promote discussion regarding what biosemiotics' ongoing implications are and how we might best represent them to fields contiguous to, and far beyond, our own. Because biosemiotics is peopled by a truly transdisciplinary section of scholars, from both the sciences and the non-sciences, there are bound to be some differences in our self-conception as a group. Some biosemioticians are devoted to the task of introducing a greater sense of the phenomenon of 'meaning' in biology. Others see the task of biosemiotics as widening the scope of an already broad church, semiotics, so that it does not mistakenly pursue the semiosis of human animals as divorced from that of other organisms. Yet another is to erode, or at the very least, to make more porous, the boundary between living nature and culture, the sciences and the humanities. As such, it is a challenge to the view of humans as 'exceptional' in nature. The second two largely represent the project in the current volume, although establishing the serious investigation of meaning in nature, through the sciences, would no doubt hasten this project; at the very least, it would assist it. In short, this book is devoted to illuminating the extent to which biosemiotics constitutes an "epistemological break" (Althusser 1969) with 'modern' modes of conceptualizing the world (including the ultra-modern posing of so-called 'postmodernism' – see Deely 2003, 2009a). It shows biosemiotics to be a significant departure from those modes of thought that neglect to acknowledge continuity across nature, modes which install culture at the centre of their deliberations and can only produce an understanding of culture which reflects that first move. This is the general cultural implication of biosemiotics. And it should be added that culture is here defined in the broadest possible way, as constituted by the practices in a whole way of life (Cobley 2008), including such routines as eating habits, table manners, sport, exercise, washing, water storage, vestments, architecture, lighting design and so forth. As will be seen, however, biosemiotics does not simply shed light on the ritual overlays on cultural practices that seem to be rooted in physical needs; it has some particular implications for those areas of culture where there is 'purely' aesthetic behaviour, apparently devoted to no survival aim whatsoever, such as storytelling, decoration, music and sculpture.

Biosemiotics offers the prospect of a renewed cultural analysis by dint of its steering a path between over-interpretation and reductionism. Some phenomena that are 'dead' for physics are evaluated in biosemiotics as embodying sign processes. At one end of the spectrum of biosemiotics are those understandings which emphasize the fluidity and growth of semiosis, attributing various degrees of agency to the most lowly of interpretants. At the other end are those understandings which identify the action of more rigid codes in nature. *Off* the spectrum at respective ends are, on the one hand, 'New Age' visions which recognize no thresholds of semiosis and see in all natural entities the presence of 'intelligence' or, worse, 'god'; and, on the other, the view dictated by absolute mechanism and Laplacean determinism in nature. In the sphere of culture, there is a series of binaries that biosemiotics abolishes or modifies by treating life as continuous and by discerning semiosis across the realm of nature; namely: individual/collectivity, agent/subject, verbal/nonverbal, human/non-human, mind/matter, culture/living nature.

The issues of continuity of matter and mind, as well as the spurious separation of nature and culture, have had, in the scheme of things, very little purchase in cultural analysis. These big issues for science have simply not translated well in the terms of the humanities. One reason for this, of course, is the way that science has offered all manner of hostages to fortune to cultural studies, from social Darwinism, through Lysenkoism, eugenics, sociobiology and the development of the nuclear bomb, not to mention science's masculine bias and other institutional factors that have vitiated its claims to knowledge. As such, observations including Kuhn's (1970) on the philosophy of science or Lyotard's (1984) report on knowledge pronouncing an era of incredulity towards the grand narrative of scientific progress, have become the common sense of the humanities. Yet there seems to be a chance, greater than any offered hitherto, to bridge the gap between the two cultures through semiotics as it has been reinvigorated by biosemiotics. Certainly Sebeok (2000) felt this.

The roots of the opportunity to bridge the two cultures can be found in general semiotics. Semiotics' levelling of the cultural playing field which provided the impetus to investigate semiosis (initially, only potentially) across all realms of life reaches its full fruition in biosemiotics. On the one hand, biosemiotics is the name given to a particular area of the entire venture of semiotics. On the other hand, the impossibility of escaping nature entails that all semiotics, however focused it is on cultural material and however much it attempts to bracket nature and cosmological considerations, is biosemiotics. A flavour of this relation between semiotics and biosemiotics is offered by Sebeok's (1986a: 60) statement – a statement that dictates the argument of the present book – to the effect that "A human body is thus an inextricably complex text that has been encoded and determined by the combined action of nature and nurture (or that minuscule segment of nature some anthropologists grandly compartmentalize as culture)". A more uncompromising formulation on the matter is difficult to imagine. Biosemiotics has also made it easier to draw together the concerns of the sciences and those of the disciplines concerned with culture by providing an approach that is more 'culture-friendly'. As Deacon (2012a: 541) remarks,

> It's time to recognize that there is room for meaning, purpose, and value in the fabric of physical explanations, because these phenomena effectively occupy the absences that differentiate and interrelate the world that is physically present.

While the study of culture continues under the impression that the natural world and the sciences devoted to studying it are geared to completely different realities from culture, then that study may be doomed to an eternal loop. Biosemiotics promises a means to interrupt that loop.

In light of this massive task, it should be noted that the current volume has a modest purpose. It attempts only to amplify some aspects of biosemiotics and to present a view of some cultural implications. The picture of biosemiotics offered here is necessarily limited and is not meant to constitute a comprehensive survey (for the closest approximation of that, see Favareau 2010a). There is, for example, no discussion of von Baer, Baldwin, Bateson or Rothschild in respect of one end of biosemiotics' history; nor is there due consideration of the major contemporary endeavours of Sharov, Pattee, Markoš et al, at the other end. Moreover, by no means all biosemiotic ideas are covered in this book. Space and focus dictate that consciousness, genes, function, need, the *Wirkzeichen/Merkzeichen* nexus and distributed language, for example, are not discussed. Nor are the constant debates in biosemiotics – for there are many – represented here, apart from in the discussion of code and interpretation. Nevertheless, I have stood on the shoulders of giants in order to gain even my limited purview in the service of identifying cultural implications. To some extent, this book merely offers a series of footnotes to points "clearly and radically" (Kull 2007: 15) stated by Jesper Hoffmeyer as long ago as 1996. It also draws heavily on arguments put forward by Kull, Deely, Petrilli and von

Uexküll, while relying heavily on the insights of Deacon; above all, this volume is guided by the work of biosemiotics' consolidator, Sebeok. It should be noted, too, that while the current volume aims to shed light on culture from the angle of biosemiotics, already the likes of Sebeok, Hoffmeyer and Deacon are themselves no slouches in providing insights for the understanding of culture through their superlative pellucid communication of complex ideas. Biosemiotics' cultural implications can be found separately in their work, too.

The present volume, then, draws together, in correspondence with its chapters, the following implications:

Implication 1: Potentially, this is the age of biosemiotics. There is now a consolidated and focused literature in the field.

Implication 2: Semiotics holds the key to understanding culture, but semiotics' project is most fully realized on a biosemiotic basis.

Implication 3: Humans are certainly 'special', but they are neither simply 'different in kind' from the rest of nature or 'different in degree'. Humans' modelling explains the foundations of culture.

Implication 4: The human's agency is not unique in the natural world. The human is a natural subject.

Implication 5: While ethics might be sustained in the short-term by a willed programme, ethics is a natural phenomenon arising out of human modelling.

Implication 6: The idea of 'codes' is a human invention. If codes occur in nature, they do not behave as they do in cryptography.

Implication 7: Humans are subject to constraints. The nature of these constraints shapes human evolution but can curb some freedoms while producing specific cultural results.

Implication 8: The arts and the humanities are natural and indispensable to the process of expanding all human experience and knowing.

This last implication, regarding 'knowing', should be taken as central to all of the argument in this book. Biosemiotics does not propose for one instant to subsume the richness of culture into a series of natural mechanisms; for the simple fact is that biosemiotics does not characterize nature as mechanical. Importantly, biosemiotics investigates how organisms 'know' their world. As will be seen, the branch of biosemiotics named 'cybersemiotics' has, in particular, attempted to theorise this 'knowing' and affords considerable credibility in 'knowing' to aspects of culture rather than just the sciences.

Of course, it is possible that the themes of this volume centre on a set of arguments which, globally, have less purchase than they do locally. This book is written from a resolutely Western perspective. In that perspective, humanism, liberalism and Eurocentrism have been powerful in establishing an understanding of culture as divorced from nature and indeed, with culture as primary, the wellspring for what is natural. Moreover, in the Abrahamic religions of the West, nature is cast in the service of God and humans. Most recently, Siedentop (2015) has argued that Western history has been seen in terms of the long instatement of moral beliefs, ultimately with the individual in the organizing social role. Secular liberalism already had its forerunner in medieval questioning of the relation of church and state prior to the Renaissance. What Siedentop shows is that individual liberty became a fundamental *natural* right in Europe, enforced by a number of judicial and ideological measures as insurance. Grounding the individual in a putative state of affairs of *nature* no doubt contributed to the shoring up of culture and social life against any need for protracted contemplation of humans' provenance in a wider conception of nature. Thus, it might be that the 'anti-humanist' perspective which this book identifies in biosemiotics may actually be a mere figment, the necessity of which is created by the specific context of the West's insulation of culture, in the guise of the individual, from the demands of wider nature. Certainly, the opposition 'culture/nature' is an English construct which does not necessarily obtain in the same way in other languages. Whether biosemiotics' abolition of the opposition can produce a sound basis for identifying demonstrable universals, distinct from the universals posited in Eurocentrism and colonialism, is a possibility to be pursued outside the pages of this book.

What follows, then, is eight chapters and a conclusion. Chapter 1 serves as a much truncated literature review. However, as it declares at the outset, the literature of biosemiotics has grown so large and covers so many complex inter-related and interdisciplinary perspectives, that it is only possible here to draw out from the literature a narrative which serves the purpose of this volume. Chapter 2 offers a sense of the place of biosemiotics within general semiotics. Because the institutionalisation of thought is often uneven, contradictory and sometimes promotes confusion, the chapter seeks to unravel for the reader some of the tangled connections in sign study.

Chapter 3 discusses how human modelling is essential to biosemiotics' answering of Darwin's question regarding whether humans are different by degree or in kind from other animals. In particular, it discusses how biosemiotics reconceptualises the nature of language and how it effectively abolishes exceptionalism. Chapter 4 has related human concerns and presents the ways in which agency and subjectivity, learning, surroundings and otherness are figured by biosemiotics.

Much of biosemiotics' implication for culture pertains to the possibilities inherent in semiosis and the constraints which allow crystallisation of semiosis into more or less stable phenomena. Chapter 5 is concerned with a very human possibility: ethics. It argues that the customary conception of ethics as a willed programme overlooks some salient problems and that, in light of biosemiotics, ethics can be seen to derive from 'involuntary' projections. Chapter 6 is concerned with the constraining power of invariance in semiosis. It tracks the concept of 'code' and discusses the character of invariance in the idea of 'organic codes'. Chapter 7 continues on the theme of constraining factors and considers the merits of the concepts of 'repression' and 'constraint' in respect of what gets left out or passed over in the dynamism of semiosis across nature. Finally, Chap. 8 discusses the cognitive, modelling drive of the arts and humanities, finding them crucial to the maintenance of human experience, the preservation of memory and the enhanced 'knowing' of the world.

Chapter 1
The Age of Biosemiotics

This chapter seeks to provide a sense of the literature of biosemiotics and to offer some orientation regarding some of the main issues that arise in this volume. With respect to the first of these two aims, it is no longer possible to write an extensive overview in the way that it might have been even just a few years ago. Not only has the literature of biosemiotics grown with contemporary publications, including those in the flagship journal, *Biosemiotics*; the literature has also grown with reference to work published in the past which is being recognized as absolutely germane to the biosemiotic project. The obvious example is the work of von Uexküll, among those who died before biosemiotics came to its present-day fruition; but there is also a great deal of work in cognitive science and in systems theory as well as in science in general that biosemiotic writings continue to invoke. With the publication of the *Semiotica* special issue on von Uexküll in 2001a (see also Barbieri 2002) and Barbieri's collection, *Introduction to Biosemiotics*: *The New Biological Synthesis* in 2007a, the question has been raised as to whether biosemiotics has 'come of age'. Arguably, by the time of the first annual 'Gatherings in Biosemiotics' conference in 2000 biosemiotics was already mature, as evidenced by the following key volumes which had built on the works of proto-semioticians such as von Baer and von Uexküll and early biosemioticians such as Prodi:

Sebeok and Umiker-Sebeok eds. *The Biosemiotic Web 1991* (1992)
Emmeche, *The Garden in the Machine* (1994)
Hoffmeyer, *Signs of Meaning in the Universe* (1996)
Deacon, *The Symbolic Species* (1997)
Sebeok, Hoffmeyer and Emmeche eds., *Biosemiotica I* and *Biosemiotica II* (1999)
Kull ed., *Jakob von Uexküll*: *A Paradigm for Biology and Semiotics* (2001)
Markoš, *Readers of the Book of Life* (2002)

This constitutes an already diverse corpus characterized by differing perspectives on biosemiotics from Copenhagen, Tartu and Prague, as well as different themes in monographs ranging from artificial intelligence to the origins of language. Added to these in the years that followed were a number of very different volumes which further diversified biosemiotics:

P. Cobley, *Cultural Implications of Biosemiotics*, Biosemiotics 15,
DOI 10.1007/978-94-024-0858-4_1

Barbieri, *The Organic Codes: An Introduction To Semantic Biology* (2003)
Weber and Depew eds., *Evolution and Learning: The Baldwin Effect Reconsidered* (2003)
Barbieri ed., *Introduction to Biosemiotics: The New Biological Synthesis* (2007a)
Barbieri ed., *The Codes of Life: The Rules of Macroevolution* (2007b)
Hoffmeyer ed., *A Legacy for Living Systems: Gregory Bateson as a Precursor to Biosemiotics* (2008b)
Brier, *Cybersemiotics: Why Information is not Enough!* (2008a)
Hoffmeyer, *Biosemiotics: An Examination into the Signs of Life and the Life of Signs* (2008a)
Markoš et al, *Life as its Own Designer: Darwin's Origin and Western Thought* (2009)
Favareau, *Essentials of Biosemiotics* (2010b)
Deacon, *Incomplete Nature* (2012a)
Schilhab et al, *The Symbolic Species Evolved* (2012)
Romanini and Fernandez, *Peirce and Biosemiotics* (2014)
Brentari, *Jakob von Uexkull* (2015)

as well as other books in the Springer *Biosemiotics* series of which the current book is a part, articles in the *Biosemiotics* journal and, now, an increasing amount of writing beyond these spheres. In addition, the annual Gatherings in Biosemiotics (see Rattasepp and Bennett 2012) continue as a focused forum for current research in the field.

Given the difficulties of summarising this work, the present chapter will restrict itself to issues which impinge on the arguments of the rest of the book. It will attempt to address questions arising from one of the founding questions of biosemiotics: "How could natural history become cultural history?" (Hoffmeyer 1996: viii). Arguably, this question guided biosemiotics in its formative years and it drives the current volume. Certainly, it is the question that lies behind extended implementation of such concepts in biosemiotics as semiotic freedom, semiotic niche, agency in nature, scaffolding, interpretation, *Umwelt* and semiosis. One quote best sums up, in a short space, the connections between these concepts. It is offered by Hoffmeyer (2010a: 34) when he describes the way in which early evolution on this planet saw the development of systems with increasing degrees of predictability that ultimately enabled them to proliferate in greater numbers. This proliferation was partly due to the ability of organisms to recognize regularities and to anticipate when nourishment might be available:

> At first such anticipatory activities would have played out at a very simple level, as when a bacterium 'chooses' to swim upstream in a gradient of nourishment rather than tumbling around waiting for the nutrients to reach it, but little by little the advantages of this talent for anticipation would have favoured any improvement of the talent that might accidentally appear, and thus would have started the ongoing tendency of evolution to create systems with ever more semiotic freedom or *interpretance*, defined as the capacity of a system (a cell, organism, species etc.) to distinguish relevant sensible parameters in its surroundings or its own interior states and use them to produce signification and meaning. An increase in semiotic freedom implies an increased capacity for responding to a variety of signs through the formation of (locally) 'meaningful' interpretants. Since semiotic freedom allows a system to 'read' many sorts of 'cues' in the surroundings it will tend to have beneficial effects upon fitness.

Here it is possible to witness the formative years of processes – *types* of activities or examples of semioses – that would spread out over time. There is anticipation, a primitive activity in this instance but one which became crucial in a very specific form in humans. There is 'choice', implying both degrees of freedom as well as the agency to make that choice. There is recognition – inaugurating a basic form of meaning, with some phenomena being distinct from surrounding phenomena. There is evolution through learning as anticipation is rewarded. There is the regularity or invariance of habit ("tendency") as it forms in nature. There is 'interpretance', the capacity to respond to signs. There is reference to semiosis inside and outside an organism. In respect of the outside, there is the reference to the surroundings as 'meaningful'. There is the depiction of a semiotic niche. There is discussion of 'fitness'. These issues will arise more than once in the discussions that appear in the following chapters. Above all, though, there is the general point that characterizes Hoffmeyer's description here: continuity of semiosis across nature.

In biosemiotics, the idea of continuity corresponds with Peirce's synechism. Peirce held that a failure to accept synechism usually goes hand-in-hand with scientific infallibilism. Scientific infallibilists are committed to discontinuity because it enables them to ascertain quantities merely by way of measurement: "For where there is continuity, the exact ascertainment of real quantities is too obviously impossible" (1.172). For the infallibilist, there are quantities not yet ascertained that may be ascertained later, plus those absolutely that are unascertainable (1.172). Put another way, a binary of measurable and immeasurable quantities is set up where there is no embrace of continuity. Precisely such binaries deriving from the denial or failure to broach continuity are seen in cultural analysis, too, and some of those are considered in relation to their undermining by biosemiotics in Chap. 4, below. Esposito (n.d.) sums up binarism and continuity thus:

> Synechism, as a metaphysical theory, is the view that the universe exists as a continuous whole of all of its parts, with no part being fully separate, determined or determinate, and continues to increase in complexity and connectedness through semiosis and the operation of an irreducible and ubiquitous power of relational generality to mediate and unify substrates. As a research program, synechism is a scientific maxim to seek continuities where discontinuities are thought to be permanent and to seek semiotic relations where only dyadic relations are thought to exist. Synechism and pragmatism mutually support each other: synechism provides a theoretical rationale for pragmatism, while use of the pragmatic maxim to identify conceivable consequences of experimental activity enriches the content of the theory by revealing and creating relationships.

Synechism, as Peirce (7.570) attested, "can never abide dualism". Dualism "performs its analyses with an axe, leaving as the ultimate elements, unrelated chunks of being". As Peirce adds (5.570), synechism holds that physical and psychical phenomena are by no means entirely distinct and although some are more mental and spontaneous and some are more material and regular "all alike present that mixture of freedom and constraint, which allows them to be, nay, makes them to be teleological, or purposive". Peirce and biosemiotics are thus aligned in seeing phenomena as continuous with qualified regularities and that that continuity inheres in semiosis. As Esposito (n.d.) adds, "Without a universe capable of expressing relational generality, signs would not exist. But signs do exist, and therefore relational generality is a character of our universe". Furthermore, the point of synechism is not simply to discern specific laws but to discern law-governance as a whole. As such, synechism is a principle which is very much imbricated with the Peircean category of Thirdness (see below), which itself accepts and nests the category concerned with regularities and seemingly discrete units, Secondness. Most importantly, synechism does not countenance distinctions of psychic and physical, self and other, plus, critically, nature and culture, except in those instances where semiosis takes on the guise of invariance, in particular habits.

Of course, the perennial obstacle to a synechist perspective which prompts the resort to the dualist axe is the problem of how matter could become mind (Delbrück 1986). Peirce notes the intractability of both physics and evolution: "Mechanical causation, if absolute, leaves nothing for consciousness to do in the world of matter; and if the world of mind is merely a transcript of that of matter, there is nothing for consciousness to do even in the mental realm" (6.613). Yet he adds (6.613) that even mechanical action involves mind of some sort and that it is possible to glimpse mind when invariance subsides

> Supposing matter to be but mind under the slavery of inveterate habit, the law of mind still applies to it. According to that law, consciousness subsides as habit becomes established, and is excited again at the breaking up of habit. But the highest quality of mind involves a great readiness to take habits, and a great readiness to lose them; and this implies a degree of feeling neither very intense nor very feeble.

The problem cannot be addressed in aphorisms, of course, but only in a grand theory. Most recently, this latter has been provided by Deacon (2012a) in a complicated, but admirably clear-headed, book.

In *Incomplete Nature*, Deacon (2012a) first addresses the matter-to-mind problem through opening chapters in which he progressively hoists on their own petard a series of 'ghost-in-the-machine' eternal regresses of mind such as the idea of homunculi or golems. This endeavour is particularly apposite in relation to that form of cultural analysis whose representatives (e.g. Adler 1967) have frequently relied on such ghosts to bolster the contention that humans are 'different in kind' (see Chap. 3, below). Like 'turtles all the way down', the idea sidesteps explanatory principles in favour of a mystical origin for cognition. Driving Deacon's thesis is the contention that there is 'incompleteness' at the heart of nature and that much of nature is populated not by chunked material but by "ententional" phenomena that are related to extrinsic processes or attempts to achieve something. An obvious example of an ententional phenomenon is 'information' and ententionality is also characteristic of actions such as those that amount to 'agency'. For Deacon, those attempts to understand consciousness with reference to models of computation are woefully far of the mark, just as much as Peirce argues that understanding the evolution of matter without factoring mind is a mistake. The existence of ententional phenomena indicates that the search for the matter-mind relation should proceed from what is "not there". Natural selection is "not there"; nor is "interpretation". What is "there" is determined by characteristic kinds of 'constraint' which work, in particular, to exercise restrictions, exclusions and thereby institute regularity or redundancy in areas of nature (see Chap. 7, below). Like the Peircean term 'habit', which has a longer lineage in biosemiotics, 'constraints' are central to evolution, they introduce invariance in nature and they differ from the absolutism of 'laws'. Peirce (6.101) writes of habits

> In view of the principle of continuity, the supreme guide in framing philosophical hypotheses, we must, under this theory, regard matter as mind whose habits have become fixed so as to lose the powers of forming them and losing them, while mind is to be regarded as a chemical genus of extreme complexity and instability. It has acquired in a remarkable degree a habit of taking and laying aside habits. The fundamental divergences from law must here be most extraordinarily high, although probably very far indeed from attaining any directly observable magnitude. But their effect is to cause the laws of mind to be themselves of so fluid a character as to simulate divergences from law. All this, according to the writer, constitutes a hypothesis capable of being tested by experiment.

Deacon (2012a: 183) notes Peirce's characterization of habits as begetting habits, which is, of course, of a piece with Peirce's perspective on mind and semiosis as continuous across nature, a perspective that is at the core of biosemiotics. But Deacon introduces the idea of 'constraint' to give 'habit' a more precise physical characterization.

One of the key issues in Deacon's 'revision' of habit is his attention to what is ententional or absent in a habit – abstract nouns related to both physical processes such as hunger and mental processes such as beauty. These are a matter, once more, of what is "not there". He then considers this with reference to: 'morphodynamics', morphological actions in nature that include self-organization and therefore regularity; and 'teleodynamics', the impetus or agency present in organisms. Teleodynamics works with morphodynamics to create a 'self'; 'dynamic reciprocity' of the internal and external worlds, including such developments as protective encasement and selective permeability (2012a: 471) are key to this action. As Deacon (2011) suggests, teleodynamics "forms a bridge from matter to what matters". In the case of a self and other, then, the constraints are not just barriers but progenitors of possibilities; they "can become their own causes" (2011), capable of maintaining – remembering and regenerating – themselves. This class of self-creating teleodynamic systems he calls the "autogen" (2012a: 307). The autogen is important because, as Deacon (2011) notes,

> the origins of life and the origins of consciousness both depend on the emergence of self: the organisational core of both is a form of self-creating, self-sustaining, constraint-generating process.
>
> Ultimately, this kind of reciprocal, self-organising logic (but embodied in neural signal dynamics) must form the core of the conscious self. Conceiving of neuronal processes in emergent dynamical terms allows us to reframe many aspects of mental life. It suggests, for example, that the experience of emotion is intimately connected with the role metabolism plays in regulating the self-organising dynamics of the brain's information-generation processes. This is because self-organised processes are generated by incessantly perturbing a system away from its equilibrium.

He adds that fMRI and PET-scan imagery might be providing evidence of conscious arousal that is not located in one place but shifting from region to region according to availability of energy in the brain.

This summary does no justice to the detail, progression and sweep of Deacon's account, an account which must be read for itself. Indeed, any summary of the argument presents a series of jumps and *Incomplete Nature* is resolutely not about anything but the most measured and nested processes. Nevertheless, the point to be made here concerns continuity and the fact that biosemiotics conceives of the matter-to-mind challenge as being answerable within accounts not dissimilar to Deacon's. Native to those accounts is the full acknowledgment of semiosis within the bodies of organisms as crucial to continuity. Endosemiosis, semiosis within bodies, is an integral part of the consideration of the relations of matter and mind. Sebeok (2001a: 19) offers a sense of the term's lineage and associations:

> The substantive "endosemiotics" was coined by Sebeok (1976, 3; see also Sebeok 1991a, ch. 1, part iii). As a consequence of Jakob von Uexküll's consistent and elaborate doctrine of signs (Jerison 1986, 143–144; Sebeok 1989d, ch. 10), nothing exists for any organism outside its bubble-like private *Umwelt* (environment) into which, although impalpably to any outside observer, it remains, as it were, inextricably sealed. The behavior of an organism – behavior being definable as the commerce by means of signs among different *Umwelten* - has as its basic function the production of nonverbal signs for communication, and first of all for communication of that organism with itself. It follows that the primal universal sign-relation in the ontogeny of an organism is realized as an opposition between the self (ego) and the other (alter); cf. Sebeok 1989d. This elementary binary split subsequently brings to pass the second semiosic dimension, that of inside vs. outside. It is this secondary opposition that enables an organism to "behave", i.e., to enter into relations to link up with other living systems in its surrounding ecosystem.

Not only is endosemiosis, as a concept, less common than it ought to be, but its role in the formation of the self that Deacon shows to be indispensable in the growth of life struggles to get on the agenda of considerations of either sociality or cultural production. Sebeok (2001a: 15), by contrast, considers what we do and do not know about the human body as an indicator of the importance of semiosis:

> Internal communication takes place by means of chemical, thermal, mechanical, and electrical sign operations, or semiosis, consisting of unimaginably busy trafficking. Take as an example a single human body, which consists of some 25 trillion cells, or about 2000 times the number of living earthlings, and consider further that these cells have direct or indirect connections with one another through messages delivered by signs in diverse modalities. The sheer density of such transactions is staggering. Only a minuscule fraction is known to us, let alone understood. Interior messages include information about the significance of one somatic scheme for all of the others, for each overall control grid (such as the immune system), and for the entire integrative regulatory circuitry, especially the brain.

In synechism, this semiosis, detected and undetected, cannot be divorced either from the higher processes of the human *Umwelt* or such experiences as subjectivity.

Possibly the key concept recurring in the current volume is that of *Umwelt* from Jakob von Uexküll. There are a number of translations of Uexküll's writings in English (von Uexküll 1992, 2001a, b, 2010) as well as a growing number of very good expositions of his work (Deely 2009b; the essays in Kull 2001; Brentari 2015). So, this is not the place to give a comprehensive overview of the concept. However, it is necessary to provide a brief outline here because subsequent chapters use the word *Umwelt* without defining it each time and, while logically following on from the consideration of endosemiosis, it is also needed for understanding the status of language in biosemiotics. For Sebeok, the closest English version of *Umwelt* is the word 'model': "All organisms communicate by use of models (*umwelts*, or self-worlds, each according to its species-specific sense organs), from the simplest representations of maneuvers of approach and withdrawal to the most sophisticated cosmic theories of Newton and Einstein" (Sebeok 2001a: 21–2). *Umwelt*, then, is the means by which organisms "capture 'external reality'" in response to semioses. Most importantly, though, an *Umwelt* is composed by the circulation and receiving, insofar as it is physically allowed by an organism's sensorium, of signs. The senses are imperative; von Uexküll (2001a: 107) considers humans:

> Around us is a protective wall of senses that gets denser and denser. Outward from the body, the senses of touch, smell, hearing and sight enfold man like four envelopes of an increasingly sheer garment.
>
> This island of the senses, that wraps every man like a garment, we call his Umwelt. It separates into distinct sensory spheres, that become manifest one after the other at the approach of an object. For man, all distant objects are sight-objects only, when they come closer they become hearing-objects, then smell-objects and finally touch-objects as well. Finally, objects can be taken into the mouth and be made taste-objects.

An *Umwelt* rests precisely on the undeniably bodily phenomena of species' sensoria. Thus, the *Umwelt* of the dog, partly derived from its acute ability to hear high-pitched sounds, differs qualitatively from that of the human whose hearing is focused on a lower pitch. What both dog and human *Umwelten* might be unfortunate enough to *fail* to apprehend poses a potential – albeit limited – threat to their survival; what dog and human share – for example, hunger, albeit in different physiological and cognitive configurations – stresses continuity in respect of life forms on planet Earth. Any species member thus 'inhabits' their *Umwelt*. A stark observation on the precariousness of life in an *Umwelt*, coupled with the evolutionary 'efficiency' of *Umwelten* is offered by Hoffmeyer (2008a: 200). A moth, he notes,

> is equipped with a totally silent Umwelt, apart from the narrow chink that is kept open for registering the bat's fateful frequencies of approximately 20,000 Hz. When the bat is far away, the moth naturally veers away from the sounds, but when the bat comes up close, the moth instead makes sudden and unpredictable movements. The moth, in other words, displays Umwelt-controlled behaviour.

The human *Umwelt* clearly does not afford humans the ability to detect the presence of bats with such a high degree of accuracy. Yet, the human *Umwelt* has other remarkable attributes.

The key point about the human *Umwelt* is that it is intricate and varied in comparison to other animals. The suite of senses possessed by humans, while it might be surpassed by some animals in highly specialised areas, as a whole offers much more sophistication than any other extant animal. Biosemiotics finds that the advantages of possessing a sophisticated *Umwelt* are numerous but, as Hoffmeyer points out (1996: 58), one of the chief ones, tied up with semiotic freedom, is *anticipation*. Another way to put this is that the concept of *Umwelt* facilitates some understanding of species' worlds yet, in the case of humans, allows the investigation of the cultural propensity for projecting *possible* worlds: fictional projections and ethical projections as well as those associated with logic, science and Thirdness and the ability to make educated guesses, or as Peirce would say, 'abductions'. In addition to opening up the understanding of realms of artistic endeavour and humane planning, the concept of *Umwelt* also offers the possibility of insight into a number of more local facets of perception and thinking. For example, Hoffmeyer (1996: 117) notes in relation to the human *Umwelt* that consciousness switches on and off relevant parts of the brain. It is easy to see how that would be of benefit in forgetting distress and remembering positivity (Sedikides and Green 2004); yet it is also probably integral to enjoying cultural artefacts. While it is a commonplace of cultural analysis that the act of representation – in whatever form, the map not being the territory – necessarily involves selection, the mystery of why humans often seem willing to tolerate and even enjoy mere representations' degrees of unrepresentativeness or why they suspend disbelief, has seldom been systematically addressed in cultural analysis beyond discussions about 'realism'. Instead, it has been relegated among the objects of speculative aesthetics.

One other issue in relation to the concept of *Umwelt* should be mentioned before this chapter proceeds, because that issue carries with it the adjunct that beyond species' capacities of semiosis there is a world – the 'real world', in one sense – which cannot be reached. Clearly, this is a philosophical matter; nevertheless, it is present and requires a note of clarification Within a species' *Umwelt* there are all manner of possibilities of 'illusion' – through misinterpretation of signs, through overlooking of signs and through signs not being 100 % adequate representations of reality. Yet, as Sebeok repeatedly pointed out (see, for example, 1986a: 14), usually referring the point back to Francois Jacob, the testimony that an *Umwelt* is a fairly good guide to reality – a workably accurate *model* – is offered by the survival of the species within a given *Umwelt*. If an *Umwelt* offered an irredeemably faulty grasp of reality, then that species would not survive. Although the philosophical issue cannot be the focus of this volume, it still appears to be a pretty strong argument for realism in biosemiotics and informs some of the exposition which follows.

The staggering capacity to differentiate objects in the world, supplemented with an ability to imagine new objects, including fictional ones, which characterizes the human *Umwelt*, is aligned with one of the human *Umwelt*'s main components: verbal language's recursive potential. This topic will be taken up especially in Chap. 3. Before that, however, it is necessary to discuss a component of that topic, 'symbolic reference', both because it recurs in the volume and because it demands some clarication. Among biosemiotic thinkers, Deacon (1997; cf. Csányi 1992) has brought to the forefront the real problems behind evolutionary accounts of language that rely on freak mutations over millennia to deliver a complex modelling system. First, he asks why there are no 'simple' languages (1997: 42ff); then he argues that "No innate rules, no innate general principles, no innate symbolic categories can be built by evolution" (1997: 338). Deacon suggests, therefore, that the expansion of the brain in human evolution was not the cause of symbol use or language, but was a consequence of it. Put another way, Deacon's argument is that there is a lengthy and convoluted process of emergence which the human brain and, especially, human consciousness exemplifies. Indeed, any "living organism is the end product of a very elaborately convolved history of recursive causal processes that cast a wider and wider net to capture sources of regularity and amplify them" (Deacon 2003: 305). The co-evolution of human brain and language entailed that the brain had to develop in order for language to develop, but the brain had to develop under the demands of language. The product was 'symbolic reference'.

'Symbolic reference' is a specific phenomenon. It does not simply lie on the proposition that humans use linguistic signs that are 'arbitrary' and that this distinguishes them from other forms of communication. "Arbitrariness", writes Deacon (2012b: 11),

> is a negative way of defining symbols. It basically tells us that neither likeness nor correlation are necessary. But this is inadequate, even though it is a common shorthand way of characterizing symbolic reference. All sign relationships include some degree of arbitrarity, because those attributes that are taken as the ground for the sign-object linkage can be chosen from many dimensions. Thus, anything can be treated as iconic or indexical of almost anything else depending on the interpretive process.

It is the interpretive process that assumes central importance in what will become symbolic reference. Arbitrariness and conventionality can be made available by symbolic reference – but they are not obligatory. Deacon notes (2012b: 11) that religious symbols sometimes use icons and invite esoteric abstractions. Furthermore, nested within their foregrounded symbolicity linguistic signs have other functions which allow the symbol to operate. He gives the example of a signet ring used as a seal on wax. The relation between impression and ring is iconic (in Peirce's terms, sharing qualities); the action of pressing into the wax is an index (in Peirce's terms, a sign *caused* by the presence of the ring and wax); the sign of social convention, the mark of royalty, which is the symbol, depends on the other two signs. So, "without familiarity with this entire system of relationships, these non-symbolic components remain merely icons and indices" (2012b: 13). It is the interpretive process involving understanding of the entire system of wax imprinting that is found to be critical here. Thus, interpreting symbolically is simply more complex than indexical or iconic interpretation.

This is not the end of the matter of symbolic reference as a human attribute. As Deacon notes (2012b: 16), Peirce always understood semiosis as occurring through one sign leading to another and another and so forth. In language, it is the "incredible size and complexity" (2012b: 17) of the infrastructure supporting symbolic reference that is remarkable. The corollary is that this infrastructure calls on a more concerted process of interpretation. The semiotic constraints at work in the wax imprint are as nothing compared with the set of constraints Deacon (2012b: 18–39) identifies in language:

(A) **Semiotic constraints**

1. Recursive structure (only symbols can provide non-destructive [opaque] recursion across logical types)
2. Predication structure (symbols must be bound to indices in order to refer)
3. Transitivity and embedding constraints (indexicality depends on immediate correlation and contiguity, and is transitive)
4. Quantification (symbolized indices need re-specification).
5. Constraints can be discovered pragmatically and 'guessed' prior to language feedback (because of analogies to non-linguistic iconic and indexical experiences).

(B) **Processing constraints**

6. Chunking-branching architecture (mnemonic constraint)
7. Algorithmic regularization (procedural automatization)
8. Neural substrates will vary on the basis of processing logic, not linguistic categories

(C) **Sensorimotor schemas & phylogenetic bias**

9. Standard schema/frame units (via cognitive borrowing)
10. Vocal takeover (an optimal medium for mimicry)

(D) **Communication constraints**

11. Pragmatic constraints (communication roles and discourse functions)
12. Culture-specific expectations/prohibitions (e.g. distinctive conventions of indication, ways of marking discourse perspective, prohibitions against certain kinds of expressions, etc.)

Without going through this list, it is still possible to point to some of Deacon's conclusions. First, the complex interpretative process is central to symbolic reference. Second, semiotic constraints are so extensive that it has been assumed that they must be innate when it is quite possible that they are learned, from a pre-linguistic stage onwards. The processes by which a child learns a language are "emergent from constraints that are implicit in the semiotic infrastructure of symbolic reference and interpretive processes" (2012b: 24). Symbolic reference betrays its provenance in organic evolution rather than any homology with machine computation: "So, for example, were we ever to find a way to engineer symboling minds in silicon, using electronic instead of chemical and ionic means of signal processing, we should expect some very different structures to emerge" (2012b: 32).

Deacon's work on language, and in general, remains faithful to the kind of synechistic perspective that this book is attempting to explicate as one of the main cultural implications of biosemiotics. Accompanying such a perspective, however, are some more philosophical arguments and these do need to be clarified before this volume can proceed. The arguments are to do with the realism which arises from both the concept of *Umwelt* and Peirce's account of semiosis. Predicated on a triadic theory of the sign, the latter approach represents an epochal departure in sign study.

In truth, there is a precursor to Peirce in the work of John Poinsot during the period of late Latin philosophy (see Poinsot 2013); yet Peirce's triadism is so thoroughgoing that it completely re-orients the understanding of sign typologies. Peirce identified three categories of phenomena that he labelled Firstness, Secondness and Thirdness which are all crucial to his theory of the sign. The realm of Firstness is difficult to conceive but is usually understood in terms of 'feeling'; Firstness has no relations, it is not to be thought of in opposition to another thing and it is merely a 'possibility'. It is like a musical note or a vague taste or a sense of a colour. Secondness is the realm of brute facts that arise from a relationship. It is the sense that arises when, in the process of closing a door, it is found that the door is stuck as the result of an object being in its way. The relation between pushing the door and its failure to close is an example of what makes up the realm of Secondness. Thirdness, on the other hand, is the realm of general laws. The law that heavy objects in front of pushed doors can prevent the door's closure is an example of Thirdness.

In a summary manner, Firstness is associated with the 'possible', Secondness is associated with the 'brutally factual' and Thirdness is associated with the 'virtual' (1.302; 1.356; EP 1.243). This is the framework in which, for Peirce, signs function. These signs are, similarly, a trichotomy in themselves, consisting of a Sign (or 'Representamen'); an Object (that which a sign refers to – either in the mind or in the world); and, the most difficult of the three, an Interpretant. Each of these corresponds to one of the three categories of phenomena: so, the Sign/Representamen is Firstness, the Object is Secondness and the Interpretant is Thirdness (2.228).

The Interpretant is that which the sign produces, its "significate effect" (Zeman 1973: 25): it is usually another sign and is usually – but not always – located in the mind. The advance inherent in the triadic sign is that the Interpretant does two jobs. Firstly, it sets up the sign relation: it is the establishment of a sign configuration involving Representamen and Object. When a finger (Representamen) points at something (Object), this is only a sign configuration if somebody else makes the link between the pointing digit and the something that is 'pointed to'. This making of the link is the Interpretant. If the finger pointed but was placed behind its owner's back, concealed from anyone else in that space, then there is no sign configuration however much the finger points. Put another way, no Interpretant is produced. The second feature of the Intrepretant consists in the way that any person looking at what the finger points to is bound to produce another sign (e.g. the finger points at the painting on the wall and the onlooker says: "Titian"). So the Interpretant is another Representamen, "an equivalent sign, or perhaps a more developed sign" (2.228). The fact that the Interpretant becomes in itself a sign or Representamen amounts to a sequence of an "interpretant becoming in turn a sign, and so on ad infinitum" (2.303). Eco (1976) concluded from this that the sign is constituted by a chain; equally, it could be said that the sign thus exists in a *network* of Interpretants (1.339) whose bearing is determined by prevailing circumstances. The sign is not so much suprasubjective, like a coded entity; rather, it is constituted in a fashion that renders it wholly susceptible to contextual factors.

Peirce actually embarked on a lifelong project (unfinished) to bring all signs into a typology and, late in life, estimated that there might be as many as 59,049 different sign types to be considered (8.343). These are generated by the three categories of phenomena (Firstness, Secondness, Thirdness) in relation to the three levels of the sign triad (which represent the sign form of Firstness, Secondness and Thirdness).

	1.	2.	3.
1.	Qualisign	Sinsign	Legisign
2.	Icon	Index	Symbol
3.	Rheme	Dicisign	Argument

The three columns (across the top of the table) refer to the three categories of phenomena that have been discussed (above). The three horizontal rows (labelled at the left side of the table) refer to the sign triad at the level of form (Firstness = Sign/Representamen; Secondness = Object; Thirdness = Interpretant). As the two axes interact, different kinds of signs are produced.

The first row – the level of the Sign/Representamen – produces three fundamental kinds of sign:

One which involves a sign made up of a quality signifying a quality – Qualisign;
One which is involves a sign made up of a quality signifying an existing thing or fact – a Sinsign;
And one which involves a sign made up of a quality signifying a general law – a Legisign.

The second row – the level of the Object – produces three more kinds of signs:

One which involves a sign made up of an existent sharing some character with its Object – an Icon;
One which involves a sign made up of an existent that has an actual physical connection with its Object – an Index;
And a sign made up of an existent which is related to its Object only by convention or habit – a Symbol. (Because of the act of *reference to an Object*, this is the trio of signs which is fundamental for Deacon's work).

The third row – the level of the Interpretant – produces three higher-grade signs:
One which involves a sign made up of a law signifying a possibility or concept – a Rheme;

One which involves a sign made up of a law signifying a fact – a Dicent sign;
And one which involves a sign made up of a law signifying reasoning or logic – an Argument.

It is evident how this relatively simple schema might be able to generate a large number of sign types (e.g. rhematic symbol) and combinations. Extended discussion of these sign types is not possible here (see Merrell 2000 for a good introductory survey). The point to note, however, regards the way in which the higher signs of Thirdness in particular are produced through the nesting of signs at a lower level of category. This is a basic requirement for following Deacon's account of semiosis in nature, for example. More generally, as other biosemioticians attest after Peirce and as was mentioned briefly earlier, synechism is a matter of Thirdness relations. It observes the general laws of semiosis across all realms rather than attending to individual signs or sign types. Peirce, however, did emphasize the way in which different sign types overlap and blend with each other such that their name only indicates a tendency rather than a fixed status (8.335). Possibly more important for the implications of biosemiotics discussed in this book is the functioning of signs not as either types or tokens but as *relations*.

It is in the consideration of signs as a relation – particularly involving signs, objects and things – that the fundamental connection of biosemiotics to Peircean sign theory can be found. The distinction is often overlooked but it is indispensable to biosemiotics even where it is not acknowledged. As Deely maintains,

> There are signs and there are other things besides: things which are unknown to us at the moment and perhaps for all our individual life; things which existed before us and other things which will exist after us; things which exist only as a result of our social interactions, like governments and flags; and things which exist within our round of interactions – like daytime and night – but without being produced exactly by those interactions, or at least not inasmuch as they are 'ours', i.e. springing from us in some primary sense (1994: 11).

Objects, on the other hand, are "what the things become once experienced" (1994: 11), bearing in mind also that experience takes place through a physical, sensory modality. In this sense, even such entities as unicorns or the minotaur can be considered objects *embodied* in the physical marks of a text. But Deely argues that a "thing of experience" – an object – requires more than just embodiment: the colosseum and the Arc de Triomphe preceded us and are expected to exist after us; but the point is that their existence as such is the product of *anthroposemiosis* (a compartment of *biosemiosis*). There are plenty of things – such as some metals in the earth and some things in the universe, as Deely suggests (1994: 16) – that anthroposemiosis has not yet touched.

Objects are thus sometimes identical with things and can even "present themselves 'as if' they were simply things" (1994: 18). Likewise, signs seem to be just objects of experience – the light from a candle, the scent of a rose, the shining metal of a gun; but a sign also signifies *beyond itself*. In order for it to do so, a sign must be: not just a physical thing; not just an experienced object; but experienced as "doubly related" (Deely 1994: 22), standing for something else in some respect or capacity (or, for short: in a context). To illustrate this point Deely employs the image of an iceberg's tip: to be sure, the tip protrudes into experience as an object; moreover, it is, as such, a thing; but, above all, as is known by the popular phrase, the tip is a sign that there is much more below (1994: 144). An important corollary of this, though, is that whatever is beneath the tip of the iceberg cannot be *approached* as a thing. It is possible that experience could make it an object but, even then, through the sensations it provokes, the feelings about them and its consequence, it is only available as a sign. Hence, Peirce's famous statement that "to try to peel off signs and get down to the real thing is like trying to peel an onion and get down to the onion itself" (see Brent 1993: 300 n. 84).

Things, objects and signs, then, imply certain relations to reality. That is, the division of *ens* (being), from its *primum cognitum* (first object) into *ens reale* (mind-independent being) and *ens rationis* (mind-dependent being). The division and its consequences are discussed explicitly in (Deely 2005, 2009a) where he resolves the division in respect of animals and their *Umwelten*. He writes (1981: 221),

> the analysis of sign – semiotic – provides a point of view that is superior to, that literally transcends, the traditional division of being into what is independent of the mind (*ens reale*) and what is dependent upon it (*ens rationis*), because in the sign, as in experience, both orders of being are found.

Humans live in the realm of signs. So, too, do other animals – but humans are 'the only animals capable of recognizing that there are signs (distinct from their practical recognition and use) and capable of developing a semiotic consciousness' (Deely 2005: 75). The overwhelming impediment to a semiotic consciousness has been the prominence in such thought of the Kantian idealist notion of the 'ding an sich', the entity that is unknowable. Coming immediately before Descartes and the moderns (and well before what Deely calls the *ultramoderns* or postmodernists falsely so called), Poinsot's Thomism offered the means to the realism that would fully inform a semiotic consciousness (Deely 2005: 76):

> Semiotics recovers the *ens reale* insisted upon as knowable by scholastic realism; yet, at the same time, semiotics demonstrates the objectivity of *ens rationis* in the social construction of species-specific realities among biological organisms. With this twofold accomplishment, semiotics manifests the distinctiveness of cultural reality in the human species as the *locus* where the differences between *ens reale* and *ens rationis* become knowable and distinguishable as such consequent upon the human grasp of *ens primum cognitum*.

Essential to the development of such a semiotic consciousness is the understanding of the difference between things, objects and signs and the way that they impinge on each other.

Once things, objects and signs are distinguished, the task of a doctrine of signs is to define exactly what a sign is. For Deely (1981: 120):

> relation involves three basic elements: what [the Latin thinkers] called the foundation, or *ground*, in our terms – some characteristic of an individual; the relation itself, which is over and above the individual – supra- and inter-subjective, we would say; and that *to which* the thing is related through its foundation, which they called the term or *terminus* of the relation.

For many, the whole of the sign is the act of representation: some entity standing in for some other entity from which it is different. This difference is important, but it is not the whole of the sign. What is frequently considered the sign – the 'relation' between some ground and some terminus – was discovered to be false. The real relation that constitutes the sign consists of ground, terminus and 'relation' as a triad. Furthermore, Poinsot delineates the functions of signs in relation to objects. As such, the relation of representation must differ from that of signification simply because an object can represent another and also represent itself, whereas it would be a contradiction for a sign to be a sign of itself. A sign is only a sign of something if that something is other than the sign. Lastly, Poinsot emphasized that the relation in a sign is not so much suprasubjective as contextual: in one set of circumstances the relation in a sign could be of the order of *ens reale*, in another set it could be of *ens rationis* (Deely 2004).

What this means for understanding how an *Umwelt* works for different species is of immense importance. The *Umwelt* of non-human animals is the 'objective' world in Deely's sense (2009a), because the animal encounters phenomena that can only be 'objects' – not independent phenomena in the complete fullness of their awareness-independent physical "reality" – dosed with the experience that the animal's sensorium affords. The human's possession of semiotics, the ability to use signs and the ability to recognize signs as such, means that the human can shift between signs, objects and things. In tandem with knowing that there are signs goes the possibility of understanding that objects might not capture all that constitutes the being of the thing. Humans, if they develop sufficient self-consciousness, have the potential to work out the degree of vacillation in which they are involved in shifts from mind-dependence to mind-independence in their use of signs. This self-consciousness is probably also responsible for humans' implementation of symbolic reference. Deacon (2012b: 19) notes that "non-human communication is exclusively mediated by iconic and indexical forms of reference and that [because] only human communication is symbolic it becomes clear why recursively structured communication is only present in humans". Humans' knowledge of signhood, their potential to shift from degrees of mind-dependence to degrees of mind-independence, is clearly related to the symbol's powers of reference having an icon and an index nested within it. It is also related to humans' ability to anticipate, imagine, to project new worlds and recreate experience.

Experience, here, also deserves emphasis. Given Deacon's identification of the centrality of the interpretation process in the use of icons, indices and, particularly, symbols, then the experience involved in 'meaning' is very much germane to cultural implications. "Meanings", writes Brier (2008a: 87, emphasis in the original),

> are the result of a coupling process based on joint experiences. This is an important foundation for all languages and all semiosis. Words do not carry meaning; rather, meanings are perceived on the basis of the perceiver's background experience. Percepts and words are not signals, but a perturbation whose effect depends on system cohesion.

At its most basic, 'meaning' arises from the process of recognition in an organism: its ascertaining that something is the same as something else or that something is different from something else (see Chap. 2, below). Yet, while such synechistic reasoning would be in line with biosemiotics and the argument of this volume, it leaves out a great deal. In its synechism, biosemiotics seems to be presenting a kind of 'grand theory' with a majestic sweep; on the other hand, it is attentive to the agency of organisms. So, while there is emphasis on collective processes and systemic operation across all domains, the experience that is sharply felt by individual organisms is not to be completely overlooked. By default, bourgeois humanist thought has incorporated such experience through its emphasis on the 'individual'.

In biosemiotics, such 'first-person' experience has probably been most pointedly foregrounded in 'cybersemiotics'.

The term 'first-person experience' indicates a focus on agency, but not the ideological construct of individual will, in the process of knowing and the operation of systems. Cybersemiotics (Brier 2008a), a synthesis of second-order cybernetics' concepts in biosemiotics, suggests that the potential of Firstness – the realm of the fundaments of first-person experience, affect and qualia – and its transformation in Secondness suffuse the processes of the plant and animal worlds as they are observed. It holds that Secondness is not only the fixing of relations but also the enactment of constraints. Brier notes that Firstness includes all known qualities (such as blue, hardness, sweetness) and, in their crystallization as invariants, they must be interpreted by a system that can recognize them as signs or habits or regularities. Second-order cybernetician von Foerster (1991) sees this process in terms of 'eigenvalues', consensual stabilising processes established by structural couplings of autopoietic (self-creating) systems in nature. These eigenvalues correspond at least to part of what Peirce calls the Interpretant: the (further) sign in the mind that occurs with a sign to establish the sign relation but also to move it on. Therefore, like cybersemiotics Peirce proposes an evolutionary science cognizant of potentials, finding patterns and dynamic modes or habits, "a science of the habits of evolution and the meaning they come to have for the living systems created in the process" (Brier 2008a: 274–5). Put another way, Peirce proposes a science devoted to investigating 'knowing' rather than just the production of eternal laws. This is precisely what cybersemiotics is seeking to promulgate. As Brier shows, Peircean biosemiotics offers a fuller theory of meaning and cognition than the one inherent in second-order cybernetics and autopoiesis. What second-order cybernetics offers to biosemiotics in return is a way to theorize the *systematization* of cognition whilst enhancing biosemiotics' orientation to knowing.

In general, cybersemiotics' emphasis on the constitution of systems is one prompt to the transdisciplinarity of biosemiotics, as opposed to a conception of biosemiotics as a mere 'theory of signs in biology'. Instead, biosemiotics is to be considered as a means for investigating *experience* in all realms of life, including first-person experience. It figures semiosis in nature as a matter of how forms of life are engaged in 'knowing'. That entails not only considering culture as a kind of 'knowing' that is continuous with the rest of nature. It also entails that many cultural practices of different kinds might be understood to be as meaningful as scientific practices in their attempt to apprehend the cosmos. This is a major implication for understanding culture. As will be implicit in the chapter, the roots of this implication lie in general semiotics' attempt to demystify some of the first-person experiences of cultural difference.

Chapter 2
Semiotics and Biosemiotics

Some of the cultural implications of biosemiotics are already inherent in semiotics. One of these is the 'levelling of the playing field' that semiotics effected. That is to say, in its interrogation of culture semiotics led the way in de-valorising *all* cultural artefacts, including those which have been said to have been born with, achieved or had greatness thrust upon them. Semiotics is a matter of understanding how sign systems – of all kinds – work. Originally, this endeavour was focused on culture: one of the key concepts of semiotics, invented concurrently by Roland Barthes (1977a) and Juri Lotman (1974) in the early 1960s, is 'the text' (Marrone 2014). Rather than a 'work', which indicates some higher purpose of an authorial genius, 'the text' indicates a fabric of devices designed through habitual sign use to reach a particular audience. Any collection of signs is a text and the concept was in the vanguard of the dismantling of the imaginary dividing line between so-called 'high' and popular culture. Thus, Literature (with a capital L) is still negotiating the cataclysm visited upon it by semiotics 50 years ago. For other fields and disciplines, semiotics has had similarly specific impacts. Linguistics, for example, has ceased to bury its head in the sand about 'multimodality'. For the last 30 years, media and cultural studies embraced semiotics in the limited, but persistent, form of the 'myth criticism' that Barthes abandoned by 1971. Marketing and brand management has followed suit. Biology, perhaps, is currently bracing itself for the latest reorientation that semiotics affords. Most importantly, though, for the present argument, is semiotics' part in the promotion of study across natural sign systems – including the cultural sign systems that are embedded in nature through the activities of humans.

That this massive, but simply stated, remit is sometimes difficult for the lay reader to grasp is a result of historical and institutional determinations as well as, perhaps, some of the anthropocentric bearings those determinations harbour. The term 'semiotics' is derived from a Greek root, *seme*, and was taken up by Charles Sanders Peirce, who sought to classify all types of signs in the universe. In this way, semiotics constitutes the major tradition of sign study ultimately derived from the ancient semioticians (see Sebeok 2001b). However, in Europe especially, it was the immense success and fashionable ascent of 'semiology' that initially brought the

P. Cobley, *Cultural Implications of Biosemiotics*, Biosemiotics 15,
DOI 10.1007/978-94-024-0858-4_2

possibility of broad sign study to the attention of the public and the academy in the latter half of the twentieth century. Semiology, of course, was inspired by the work of the Swiss linguist, Ferdinand de Saussure, whose *Cours de linguistique générale* (1916) predicted the growth of a general science of signs that might be possible if his principles were followed. In the latter part of the twentieth century, Saussure's call was taken up by semiologists (for example, Barthes 1973; Guiraud 1975) who confined their analyses to a limited range of cultural artefacts that might be susceptible to elucidation using broadly linguistic principles. Semiology prospered in Anglophone academia from the 1960s to the 1980s, gelling with the currency of (English) literary studies and sociology, as well as the popularity of Marxist politics.

Because of the centrality of textuality to semiotics after Lotman and Barthes, a current of thought which gained considerable traction in the humanities and the social sciences in the latter part of the twentieth century became erroneously associated with semiotics. This was the 'linguistic turn', inaugurated by Richard Rorty's 1967 influential collection, which coalesced various perspectives including those that later became prominent in Anglophone cultural studies. The idea that knowledge is 'constructed in discourse' with humans' apprehension of the world amounting to a mere figment induced by figures in language, arose out of the 'linguistic turn' and (post-)structuralism. As will be seen, the nominalism of the 'linguistic turn' is at odds with the Peircean realist perspective in biosemiotics. It also posits a definition of language based on 'figures of speech' and 'chatter' (see Chap. 3, below) rather than the more sophisticated cognitive perspective in biosemiotics offered by language as modelling.

The assumption, stemming from the linguistic turn, that much of human life was 'constructed in discourse' also underpinned efforts to conduct 'communicative praxis' (see Chap. 5, below). Barthes' programme of ideology critique launched in 1957 with his much translated work, *Mythologies*, provided an agenda for systematically analysing and rejecting the superstructural products of capitalism (Cobley 2015). The systematic aspect of Barthes' ideology critique derived from Saussure's separation of two sides of a linguistic sign into (a) a 'sound pattern' in the mind which represented sensory impressions of sound outside the mind; plus, (b) a 'concept' consisting of an abstract formulation of phenomena in the world such as 'house', 'white', 'see' and so forth (de Saussure 1983: 65ff., 101ff.). Saussure referred to these as *signifiant* and *signifie*, respectively, and the first principle regarding their connection that he emphasized was arbitrariness (1983: 67–70). Saussure's *Cours* was first translated into English in 1959 and *signifiant*, *signifié* and *signe* were rendered as 'signifier', 'signified' and 'sign'. The first item gave the impression to English natives that the *signifiant* was anything that did the work of signifying or, to put it another way, a sign – precisely the formulation that Saussure wanted to avoid. The term for the *signifié*, at the same, seemed to be anything that was the object of signification. At a stroke, Saussure's psychological conception of the sign was lost and versions of semiology were given free rein to look at all manner of cultural artefacts as if they embodied a *signifié/signifiant* relationship. The matter was compounded by the currency of Barthes' influential primer, *Elements of*

Semiology, translated into English in 1967. In order to enable semiology to be extended beyond linguistic signs, Barthes effected a slippage from Saussure, suggesting that "the signifier [*signifiant*] can, too, be relayed by a certain matter . . . the substance of the signifier is always material (sounds, objects, images)" (1967: 47). Barthes is not shy about the reasons for this un-Saussurean assertion: it was made so that the matter of all signs, including those in mixed systems, could be considered in the same way (1967: 47). Not only was there an encouragement to focus on those sign systems that were dominated by verbal modes, then, semiology also insisted that even nonverbal modes were susceptible to analysis based on the principles of Saussurean linguistics. In all cases, however, the sign systems to be analysed were human in origin.

Semiology therefore thrived in the humanities and, especially, along with 'discourse study', in established disciplines such as linguistics. It is this early institutional ascendancy of semiological principles which can often confuse the lay reader, along with the fact that the anthropocentric endeavours of semiologists were brought together with those of semioticians for the formation of the International Association for Semiotic Studies in 1969 (see Sebeok and Cobley 2010) under the banner of 'semiotics'. If semiology created the impression that the whole of sign study was human discourse and the human sign such that "All that is left is different forms and combinations of power and meaning games in a post-modern age" (Brier 2008b: 35), semiotics in the wider sense demonstrated something very different. The very localised study of the *linguistic* sign, a sign type used by humans alone, is only one component of the study of the sign in general. The human phenomenon of language is just one minuscule aspect of a broader *semiosis*, the action of signs throughout the universe no matter how they might be embodied. Put this way, language looks very small compared to the array of signs engendered by all interactions between living cells. Moreover, the issue of what is living is crucial: many semioticians of the major tradition, influenced by (Sebeok 2001c: 6), see semiosis as the "criterial attribute of life". Sebeok, building on the work of his teacher, Charles Morris, as well as the sign theory of Peirce, carved out the study of non-human semiosis originally with his work in zoosemiotics (1963). Superseding this has been a fully-fledged biosemiotics in which it is recognized that not just a semiotics of human communication is needed, but, in addition to zoosemiotics, a semiotics of plants ('phytosemiotics'), of fungi ('mycosemiotics') and of the 3.5 billion year old global prokaryotic communication network within and between different bacterial cells ('microsemiotics, cytosemiotics'). Indeed, contemporary semiotics recognizes that the human, while s/he is a sapient user of signs, is not just a discursive entity: in fact, the human is a mass of signs enacting message transfer nonverbally within the body ('endosemiosis').

It was with the advent of zoosemiotics from 1963 and, then, especially biosemiotics, that semiotics became recognizable as a pre-Socratic enterprise seeking to unify science and philosophy. That is to say, semiotics' concern became the operations of the entire cosmos – the Earth, its inhabitants and the elements – rather than just the interactions that constitute the polis. Both Peirce and Sebeok, out of step with the intellectual fashions of their times, shared this outlook. For the later Peirce, especially, the entirety of logic, philosophy and science were only approachable through an expansive sign theory, as Poinsot had demonstrated in 1632 (see Poinsot 2013, Chap. 1, above, and this chapter, below). Peirce envisaged a sign theory that would be comprehensive rather than localised, comprising "mathematics, ethics, metaphysics, gravitation, thermodynamics, optics, chemistry, comparative anatomy, astronomy, psychology, phonetics, economics, the history of science, whist, men and women, wine, metrology" (Peirce 1966: 408). It did not defy logic when he wrote to Lady Welby, late in life, revealing that he had recognized ten basic types of signs and, as has been seen in Chap. 1, 59,049 different classes of signs in all (Peirce 1966: 407).

Whether signs covered the entirety of the universe of just humans' activity, it is important to note that the history of sign theorizing from the ancient medics onwards was largely dominated by a binary distinction of *signans* (the vehicle which acted as sign) and *signatum* (that which was signified). A quasi-necessary consequence of this two-sided relationship is a 'code' perspective in which the 'vehicle' is an encoding of some content or 'tenor' (cf. Richards 1937 and this volume, Chap. 6, below). The high point of this binarism is to be found in Saussure's *Cours*: the sound pattern (*signifiant*) and the concept (*signifié*).

Saussurean semiology is not principally concerned with how signs indicate or communicate about specific objects; instead, its focus is how regimes of communication, somewhat removed from specific objects, are sustained and perpetuated. This has been a productive perspective and has spawned much work that helped in decoding the familiar and the further reaches of culture. However, the key observation is that Saussurean semiology has largely served a conception of signification as communication. It has not fared so well as a means to explicate cognition, the relationship of communication and cognition, the broader world of signs and the *Umwelten* of sign users.

Whereas Saussure continued the tradition of the two-sided sign, Peirce broke with this line of thought and insisted on a triadic sign. The theoretical importance of this break should not be underestimated and it has been emphasized and discussed in the previous chapter; yet it was by no means without precedent. Its roots can be discovered in Peirce's profound knowledge of not just classical logic but also of the Latin scholastic tradition. As with the technicalities of what was taken from Saussure's *Cours*, it is important to consider the mechanics of sign-hood inherited from this tradition. The Latins took as part of their task the exegesis of the perspective on signs emanating from the teachings of St. Thomas Aquinas. The most important of these exegetes was the aforementioned John (sometimes 'Jean' or 'Joao') Poinsot: his '*Tractatus de Signis*' (2013 [1632]), nearly 60 years before Locke coined the term 'semiotics', offers a realist foregrounding of the sign as the object of study to illuminate the two key states: mind-dependent being (*ens rationis*) and mind-independent being (*ens reale*). Deely (1994: 11–22, cf. 2009a), the scholar who rescued Poinsot from mere footnote status, demonstrates how Poinsot defined an *object* as always an *object* of experience (an entity involving mind-dependence), definitionally distinct from a *thing* (a mind-independent entity). As was seen in Chap. 1, the latter may be made an *object* by the *thing* being experienced; but, even then, through the sensations it provokes, the feelings about them and its consequence, that *thing* is never available 'in full' – it is only available through a *sign*. That the *sign* is simultaneously of the order of mind-independent *and* mind-dependent being and that it is triadic in constitution ('Representamen', an Object and an 'Interpretant') as Peirce theorized, is of great import for biosemiotics.

Biosemiotics and the Saussurean sign were not compatible, despite the attempts of individuals such as the Belgian biochemist, Marcel Florkin (1974), to unite them. Although it only became explicit a little later, Sebeok's foundation of zoosemiotics proceeded from a broadly Peircean perspective on the sign. More importantly, from the late 1970s onwards, following his discovery of the 2nd German edition of the *Theoretical Biology* of Jakob von Uexküll in 1976, Sebeok began to develop semiotics (and biosemiotics) in Uexküll's direction (see Sebeok 2001b). As has been discussed in Chap. 1 and will be revisited in Chap. 3, von Uexküll's work, even when it is not mentioned, is integral to biosemiotics, particularly the formulation of *Umwelt*. Biosemiotics takes it as *read* that all species live in an 'objective world' that is constructed out of their own signs, the latter being the result of their own sign-making and receiving capacities. In relation to general semiotics, Peirce had already stated that "A sign, or *representamen*, is something which stands to somebody for something in some respect or capacity" (2.228). It is evident, then, that there is a fit between the sign in Peirce's definition being *for* "somebody" (or some species member) and von Uexküll's idea that any animal lives in a world where the signs it circulates are characteristic of its species and sensorium. Moreover, the notion of *Umwelt* maps nicely onto Deely's Peircean formulation of thing/object/sign. Non-human animals, it should be clear, do trade in signs; however, they inhabit an objective world where their experience determines the character of what they apprehend.

Two consequences arise from these facts and should be noted here as a reminder of the features of biosemiotics that this book seeks to amplify. The first is that biosemiotics is not just a matter of explicating nature in terms of communicative signs. Instead, it charges itself with the task of understanding the 'experience' of signs that occurs in nature, how organisms 'know' the world and how the highest organisms have 'cognition'. The second is that the non-human animal's dwelling in an 'objective world' means that it cannot ponder the mind-independent being. While it is true that such animals can implement signs, unlike the human they do not know that there is such an entity as a sign which is susceptible of analysis (see Deely 2010). The animal with an *Umwelt* that facilitates such knowledge is the human. Yet, it is as well to be immediately clear that this does not entail that the human in biosemiotics is a fully autonomous entity, in a special category, divorced from nature. The human, with its recognition of signs and all its paraphernalia of culture which seems to depart at such length from the apparently lowly mechanical processes of nature, is part of a natural continuum. Indeed, the reason that the human does not depart from the mechanical processes of nature is that biosemiotics demonstrates that those processes are often actually far from mechanical. The cultural implication is that humans must be considered for their consanguinity with other living organisms whose operations, as will be discussed in Chap. 3, are less mechanical and more in tune with human semiosis than was thought before biosemiotics' insistence on investigating how organisms 'know'. Of course, at the same time it is worth considering that humans, because of the disjuncture of *Umwelten*, may not be in a position to assess the import of non-human animal semiosis. As De Waal states (2016: 6) when pondering the injustice of the *scala naturae*, "It seems highly unfair to ask if a squirrel can count to ten if counting is not really what a squirrel's life is about". The point is not incommensurable with Deely's formulation of the semiotic animal (cf. Colapietro 2016).

One way to approach what unifies biosemiotics, and to do so in a way which will reveal cultural implications, is to consider its objects, the key phenomena of its investigations. Kull (2007: 2) suggests the following list: recognition, memory, categorization, mimicry, learning and communication. Most of these would not be on the lay person's list of attributes to note in the world of living non-humans. What they demonstrate, once more, is the perspective according to which biosemiotics elucidates the continuity of nature in considering what natural entities might be considered to 'know' through their implementation of signs.

C*ommunication* most clearly bears on what anthropocentric (pre-biosemiotic) discourse understands as the division between nature and culture. All organisms communicate in some way. The difference between what is human and non-human is not to be predicated on communication qua communication; rather the question bears on what is verbal and what is nonverbal. Communication has a role to play in what constitutes an agent and a subject and what putatively separates the individual from the collective.

Learning is often associated with the experiential process which humans undergo. It is commonly observed in the activities of young animals, for example those that spend their early months in 'play' as a prelude to hunting. Yet, learning needs to be considered in a new light. "Once alive", writes Kull (2014a: 288) "organisms cannot avoid fulfilling their organic needs and, by doing so, they cannot completely avoid learning". This is because "life is a more-or-less continuous problem-solving process" (2014a: 292). Learning separates the human and the non-human by degree and by quantity and has its role in subjectivity and agency as well as in distinguishing between nature and culture. Yet, as a semiotic process, stripped to a set of structural co-ordinates, it is clearly a continuous phenomenon necessitated by life.

Mimicry is a phenomenon in the natural world which, since Aristotle observed the chameleon, has seen its semiotic features neglected or underplayed. Maran (2007) re-dresses this and considers the role of mimicry in *Umwelten*. He identifies 'abstract mimicry' "where the object of imitation is a semiotic structure with such an intense or general meaning that its connection with a particular form has obtained secondary importance" (2007: 244). Where biology has tended to understand mimicry in terms of resemblance of animals, biosemiotics identifies the semiotic process. While, mimetic features are embodied, Maran points out that they are subject to semiotic rules, "where perception, resemblance, interpretation, messages, meanings, and their later consequences become decisive" (2007; 244). The signs of mimicry are obviously crucial in subjectivity and in belonging to a collective, as well as the purpose they serve for survival.

Following Lakoff and Johnson, Kull et al. (2008: 46) note that every living thing categorizes. Moreover, this opens the question of how distinctions are made by organisms and in organisms. These are considered to be part of "the wealth of scientific questions that have been left unanswered – primarily because they have been left unasked – by the nonsemiotic life science" (2008: 46). Leaving aside the more nuanced descriptions, including those made by human animals in culture, what is harmful, what is beneficial and what is neutral or safe to ignore are central categories for the definition of culture as a 'higher process' as they are to the survival impulse of organisms other than human beings. Likewise, *memory* has served survival. Although commonly understood as a means to preserve and reproduce information, memory is a process imbricated with recognition, meaning and inheritance. So, Kull emphasizes that any semiotic system has its own memory. Moreover, organisms, as semiotic systems, co-exist with other organisms. Thus,

> All living organisms demonstrate plasticity, i.e., acclimatization to the conditions that occur. The extent of the response is, of course, very different in different groups of organisms. The particular form of plastic response is often unique. Organisms are capable of adaptive response even if the situation is completely new (in the sense that the organism has never encountered such a situation in the whole history of life). If a response becomes a habit (or a conditioned response), i.e., if it is remembered, it is called learning. Habituation is almost as universal a feature as plasticity; it occurs in all organisms as long as they are alive. Habituation means that a solution, once found, will be found easier the next time; this facilitation in repetition is due to various mechanisms, together called memory. Consequently, learning (defined as plasticity plus habituation) can be one of the attributes of life. (Kull 2014b: 52)

For biology, memory is a matter of inheritance (epigenetic, neural, and social) but semiotic processes include memory processes in general (Kull et al. 2009: 172). This applies all the way down to the cell "where the relations between the signal received and the action followed can be related to the third – for instance to the lack or excess of something in the cell that can be regulated by the appropriate action" (Kull 2010: 51). For humans, there has commonly been a distinction between individual memory (indigenous to each person) and collective memory (usually sustained by cultural heritage devices). Yet, since as long ago as Bartlett (1932), that distinction has been shown to be misplaced. When considered as a semiotic process, memory's domain and project is a key component in the networked relations of an *Umwelt*. This suggests that while material manifestations occur, it is a mistake to consider memory as solely a mental phenomenon of which the material is a representation. As Deacon (2012a: 424) notes, when discussing 'constraints', the changing distributions of electric charge in the memory registers of a computer are not the crucial elements so much as what is being transmitted.

Along with memory, *recognition* has been the process through which collectivities have been organized in culture, through which humans have been subjects and/or agents, the process by which humans have orientated themselves to others verbally or nonverbally and how both nature and culture and mind and matter have been separated. In biosemiotic terms, meaning is a unit of recognition because any organism that does something more than once is encountering meaning. Referring to Uexküll's 'functional cycle', Kull (2004: 104) notes that "all behaviour of organisms, all functions of a living body, are expressions of circular acts which include recognition of signs by receptors, actions as induced by these recognitions, and perceptions of the results of these actions". It hardly needs to be stated that 'recognition' in English signifies a 're-cognition'. What does need to be added, however, is that recognition as a semiotic process, where distinction arises from reinvesting objects, is continuous across nature.

What has prevented recognition – as well as memory, categorization, mimicry, learning and communication – being assessed as processes, is the problem of their objects over different domains. This is the translation problem quoted in Chap. 1, above, that Hoffmeyer (1996: viii) identifies at the outset of *Signs and Meaning in the Universe* when he writes “How could natural history become cultural history?” When a lowly organism carries out a recognition, it seems so different in quality from when a higher organism enacts the same genre of process. In the latter instance, ‘meaning’ seems almost immeasurably heightened in comparison to the former instance. Ameliorating this difference, requires a train of thought to the effect that, as Hoffmeyer writes (1996: viii), “something become[s] ‘someone’”. While biosemiotics is regularly observed and commended for its exposure of semiotic processes across nature, it also presents an important implication for culture by offering the reminder that human practices of meaning (through recognition, memory, categorization, mimicry, learning and communication) are not exceptional.

The real problem that has prevented this implication from becoming a commonplace is that the translation of semiosis needs to be considered in different relations to time as well as in terms of the disparity of the realms of ontogenesis and phylogenesis. Consider the short period in which culture has existed on Earth and the rapidity with which it has developed. Then consider the evolution of flora. As Nöth writes, “Unlike in human or animal communication, where a sign can be produced rapidly and its purpose interpreted immediately, evolutionary plant semiosis is a phylogenetic process in which sign production occurs in the form of evolutionary selection” (2007: 147). There is, then, a major problem of translation between biosemiotics and cultural analysis in terms of the ontology of the object of both. Yet there is at least one other translational problem in relating the two areas.

While biosemiotics has inculcated conceptions of agency and semiosis, cultural analysis has been less receptive to calls for it to contextualize the human in terms of its natural heritage. As has been argued already, the human has often been taken as an absolute exception and the analysis of culture has perpetuated itself precisely through exceptionalism (cf. Harries-Jones 2016: 2). That is, apart from the occasional social Darwinist or vulgar determinist representations of the evolution of culture, humans and their practices have been seen as overwhelmingly ‘different in kind’ from all other life on the planet (see Chap. 3, below). This is not just a throwback to Biblical or other religious narratives in which the Earth is the centre of the universe. Indeed, the problem lies in the secularist ferments of the Renaissance, where humanism strived to provide an alternative to the human as subject of the church. The compromise between the two positions can be seen in the work of later humanists such as Mortimer Adler (1967) who want to avoid the ‘ghost in the machine’, homunculi or golems (Deacon 2012a) that are attendant on Platonic or Cartesian dualism but nevertheless insist that there is some kind of ‘leap’ in evolution or some special quality that eludes evolution and entails that humans are ‘different in kind’ from other animals. Even as humans became ‘naturalized’ in seventeenth-century Western science, became the object of empirical and quantitative studies, Gaukroger (2016) argues that it was the 'moral' or human sciences which came to the fore and that "considerations of our relation to the natural realm now shape conceptions of the natural realm itself" (8). In the attendant 'humanization' of nature that was to underpin modern science, anatomy and religion, especially, were central players in "an aestheticized humanist conception of the aims and meanings of life" (309). It is thus probably true to say that most discussions of culture take place as if in an evolutionary vacuum, with only very partial thought given to the rest of the natural world in which culture is embedded or a with conscious commitment to an explicitly humanist agenda. Nor is it necessary to compel even all semiotic analyses of culture to constantly reference an evolutionary framework. However, at moments of crisis, the humanist underpinnings of some theories of culture are exposed, along with their poverty (see Chap. 8, below). Such understandings of culture still have some way to travel before they can meet even biosemiotics' more agent-friendly scientific approach to nature. It is for this reason that the present volume still argues for an anti-humanist perspective which, in spite of biosemiotics' departure from mechanism and materialism, remains necessary in assessing culture.

Physicalist science's prohibition of observations of agency in nature has been anathema to the humanities and arguably at the fundament of the 'two cultures' split (Snow 1959) or the ability to posit such a phenomenon. Biosemiotics has constituted a critical voice in this dimension of the sciences, identifying the restraining force of sterile scientism; for example, Hoffmeyer (2011: 191) has written about the counter-intuitive bent of "eliminativism" which denies "the reality of unlawfulness in the natural world, and thus of human free will". Yet the maintenance of the 'two cultures' has been largely effected by the arts' and humanities' traditional refusal to translate from the sciences or to even engage with them. This refusal has promoted an isolationist position in which humans and culture are not just a special case but are simply unreachable by any form of science when a simple acknowledgement of nature as a continuum which includes cultural practices would effectively be the first step towards abolishing the separation between 'the sciences' and all the other disciplines. This acknowledgement is embedded in biosemiotics through its adherence to the synechism that was advocated by Peirce and it is a logical consequence of general semiotics' focus on sign systems or semiosis rather than just the substrate of an individual sign. As discussed in Chap. 1, above, synechism is the principle of continuity; it is also associated with Peirce's category of Thirdness, the realm of laws or, to put it another way, the 'underlying phenomenon' which seems, at first glance, not to be a substance itself. Peirce explains,

> There is a famous saying of Parmenides {esti gar einai, méden d' ouk einai}, "being is, and not-being is nothing." This sounds plausible; yet synechism flatly denies it, declaring that being is a matter of more or less, so as to merge insensibly into nothing. How this can be appears when we consider that to say that a thing is is to say that in the upshot of intellectual progress it will attain a permanent status in the realm of ideas. Now, as no experiential question can be answered with absolute certainty, so we never can have reason to think that any given idea will either become unshakably established or be forever exploded. But to say that neither of these two events will come to pass definitively is to say that the object has an imperfect and qualified existence. Surely, no reader will suppose that this principle is

> intended to apply only to some phenomena and not to others, – only, for instance, to the little province of matter and not to the rest of the great empire of ideas. Nor must it be understood only of phenomena to the exclusion of their underlying substrates. Synechism certainly has no concern with any incognizable; but it will not admit a sharp sundering of phenomena from substrates. That which underlies a phenomenon and determines it, thereby is, itself, in a measure, a phenomenon (7.569).

The semiotic process, then, is what is at issue in nature's continuity rather than the physical being alone of any substrate. If the consequences of this for understanding culture have not been demonstrated sufficiently thus far, consider Peirce, once more, on the problem of matter and mind not as "two radically different substances but two empirically different aspects of the same substance" (Colapietro 1989: 89):

> In view of the principle of continuity, the supreme guide in framing philosophical hypotheses, we must, under this theory, regard matter as mind whose habits have become fixed so as to lose the powers of forming them and losing them, while mind is to be regarded as a chemical genus of extreme complexity and instability. It has acquired in a remarkable degree a habit of taking and laying aside habits. The fundamental divergences from law must here be most extraordinarily high, although probably very far indeed from attaining any directly observable magnitude. But their effect is to cause the laws of mind to be themselves of so fluid a character as to simulate divergences from law. (6.101)

A Colapietro (1989: 89) observes, this is an idealistic position since it makes matter a species of mind, but it is simultaneously a materialistic position because it insists upon the embodiment of mind. Hopefully, without forcing the analogy inappropriately, it is possible to see here the necessity of comprehending nature's provenance of culture. Substrates in nature might be considered as habit fixations of mind, while the laws of mind that have produced culture (as well as nature) have featured divergences in cultural practices that are themselves of so fluid a character as to simulate divergences from law.

Arguments regarding human exceptionalism thus verge on the mystical, forgetting or denying that humans and culture are subject to any physical principles at all. Humanism, as poststructuralism and postmodernism recognized albeit in a limited and self-serving fashion, is predicated on the unwarranted assumption that humans are central in the cosmos. Anti-humanist – and sometimes 'posthumanist' – scholars have eschewed individualism and cultural vitalism, attempting to depict the human as *subject* to the structures that humans have often been instrumental in constructing (see Chap. 4, below). Semiotics' insistence on neutral analyses, focusing on the 'how' of sign systems, has also tended to evacuate human values from the phenomena under scrutiny. Biosemiotics, by seeming contrast, has been committed to exploring agency in signification, sharing insights into 'autonomy' with biophysicalist complex science (see Kauffman 2000; Neimark and Ake 2002). The temptation to over-emphasize free will in light of biosemiotics is, of course, to be avoided. The same kind of over-emphasis in culture has been responsible for the idea of 'art' as absolutely autonomous. It is also the way that the arts and humanities have insulated themselves from the much wider world which science investigates. Biosemiotics offers an entrée for a revolution in the understanding of culture; but the translation problem concerning the relative weights put on agency in culture and agency in nature will have to be negotiated carefully. Seeing beyond this, it is more circumspect to identify an implication in the biosemiotic infusion of general semiotics that will be taken up in Chap. 4: that the human is constituted by nature.

Another word for the exceptionalism or the insistence on the complete autonomy of culture is 'anthropocentrism'. Unsurprisingly, in anthropocentrism the specific qualities of the human are accepted as a discontinuity with nature, often suspending evolutionary thought altogether. The key quality in this case is the capacity for language; thus, the 'linguistic turn' and the idea of the world as 'constructed in discourse' add glottocentrism to the exceptionalist mix. Glottocentrism has served well as an institutionalized comfort zone, a prophylactic against the claims of continuity in nature. In contrast to semiotics, specialization has constituted a disciplinary self-perpetuation where the demand to be accurate about a very localized phenomenon inoculates against the need to recognize that phenomenon's natural determinations. A particularly strange case of this is linguistics since 1945 which, on the one hand, has seemed to embrace its natural underpinnings with the advent of discussions about 'universal grammar' in the late 1950s and, on the other, has started to recognize in 'multimodality' that language cannot be isolated from other kinds of semiotic modelling. Yet, rather than pursuing these with a vigour that comes from unity of purpose, linguistics has split off into myriad schools which seldom if ever speak to each other. One result of this has been that 'language' has become the site of a free-for-all in which it has been co-opted to support conflicting glottocentric positions. As Hoffmeyer writes, "Ambiguous definitions of the differences between *words*, *sentences*, and *language* on the one hand, and *reference, meaning* and *understanding* on the other, has allowed too much room for metaphoric and misleading reasoning" (2008a: 281).

Biosemiotics has taken on the challenge of effecting change in science and, as such, is well aware of the problems of translation between the sciences and the humanities. The different imperatives in respect of the importance of agency in biosemiotics and the analysis of culture – the former seeks greater acknowledgment of agency while the latter is dogged by the problem of agency being overblown – is a relatively small impediment to recognizing the cultural implications of biosemiotics. Emerging approaches in environmental humanities, ecocriticism, ecophenomenology, cultural ecology, the study of embodiment, and posthumanism indicate a desire for the kind of revolution in understanding culture that biosemiotics so clearly and radically presages. The conflicts in institutionalised glottocentrism entail that some species-level issues concerning language which afford the potential benefits of a broader view remain off the agenda. In cynical institutional terms, this is understandable – if a discipline and its workers can become self-perpetuating and removed from what might be seen as the deleterious effects of other disciplines, there can be little surprise at the desire for maintenance when this is achieved. Yet, some of (bio) semiotics' nearest neighbours – and, often, most institutionally powerful potential collaborators – in the study of significance have barricaded themselves against intruders by way of specialization and anthropocentrism.

Nevertheless, it is clear that biosemiotics does owe a great deal to the 'old' semiotics and that which it owes regards an important cultural implication. In sum, it is the dedication to interrogating all kinds of sign systems without bias towards one or the other. For the semiotics of the post-Second World War period, this dedication finally overturned the hierarchy of 'high' and popular culture, a major landmark in the challenge to authority that was mounted across culture and social life, with varying degrees of success, in the last century. Yet, it is clear that either semiotics of this period was thinking too small or that the democratisation of culture that it entailed was only a short-term aim. Certainly, it is clear that the 'culture wars' that were ignited by the opening up of interpretation by semiotics were not unproblematic (Eco 1990; Dunant 1994). At the same time, the genie was out of the bottle, with semiotics' undermining of the bourgeois hierarchies of culture promising still something more. By 1971, Barthes was able to declare, in evaluating his *Mythologiques* 14 years after its publication in French, that "denunciation, demystification (demythification)" (1977b: 166) of the bourgeois and the petit bourgeois had become, itself, a mythological *doxa*. 'Mythoclasm' was to be succeeded by 'semioclasm', he claimed, a far-reaching interrogation of *all* sign systems and a *challenge* to their very basis. This would not simply entail unravelling the connection of denotation and connotation which sustained certain cultural hierarchies as 'natural', but a more thorough assault on the mechanics of meaning at the very level of the sign itself.

Barthes' call for *semioclasm* came shortly after the formation of the International Association for Semiotic Studies in 1969, where semioticians such as Thomas A. Sebeok broadened the entire agenda of sign study by encouraging its application to the whole of life. Barthes' subsequent 'retreat' into highly personalized writing, taken in this context, was not entirely without its political co-ordinates. However, the project of semiotics continues with the uncovering of sign processes throughout the living world. This is not just a matter of finding more objects for semiotics. Unsurprisingly, following the fashionable moment of semiotics in the West during the 1970s and early 1980s when semiotic analysis still had the flavour of magic, the commitment to semioclasm – even in hitherto unexplored realms for such analysis – seemed to some to be just more sterile analyses of different phenomena. In addition, it probably seemed to the casual observer that it reveals very little about humans and what impinges on them in the polis. Such a view, of course, constitutes a grave error. The implication for culture of biosemiotics' infusion into general semiotics is that analysis no longer promises to reveal simply what the messages that humans send are like: how they are constituted and structured. The commitment to considering semiosis as continuous across the realm of nature changes that imperative. If it seems that, in doing so, biosemiotics is treating immaterial phenomena, then this is not a problem for semiotics but a problem of physicalist science which, as Deacon (2012a: 23) indicates, does not deal with the content of a thought, the goal of an action or the conscious appreciation of an experience: "They aren't exactly anything physical, even though they depend on the material processes going on in brains". The same could be said for the human use of signs.

Semiotics, now casting its net to analyse sign systems in the whole of nature, is thus concerned with how humans operate amidst signs, what distinguishes their cognition and their being as endosemiotic phenomena among other organisms and in the cosmos. Put another way, all semiotics that eschews exceptionalism is biosemiotics; this, in turn, is semiotics. Answers to questions about human affairs, as will be seen in the next chapter, are sought in the interrogation of modelling.

Chapter 3
Difference in Kind or Difference of Degree?

Possibly the most striking implication for culture arising from biosemiotics concerns the answer to the old question regarding whether humans are different in kind from non-human animals or whether the difference is a matter of degree. One imagines that the argument arose less frequently in the West before 1859 when Darwin published *The Origin of Species*. By 1871, Darwin had actually phrased the issue in what has become common parlance since the publication of *The Descent of Man*, asserting that

> the difference in mind between man and the higher animals, great as it is, is certainly one of degree and not of kind. We have seen that the senses and intuitions, the various emotions and faculties, such as love, memory, attention, curiosity, imitation, reason, &c., of which man boasts, may be found in an incipient, or even sometimes in a well-developed condition, in the lower animals. They are also capable of some inherited improvement, as we see in the domestic dog compared with the wolf or jackal. If it be maintained that certain powers, such as self-consciousness, abstraction, &c., are peculiar to man, it may well be that these are the incidental results of other highly-advanced intellectual faculties; and these again are mainly the result of the continued use of a highly developed language. At what age does the new-born infant possess the power of abstraction, or become self-conscious and reflect on its own existence? We cannot answer; nor can we answer in regard to the ascending organic scale. The half-art and half-instinct of language still bears the stamp of its gradual evolution (1871: 105–6).

A few ongoing research programmes are identified in this quote, including some that claim to have made great progress. One which stands out and has become a commonplace in respect of degree/kind arguments since Humboldt is the one concerned with language. There is neither space nor time, here, to launch a detailed discussion of definitions of language; but a few points in respect of biosemiotics' implications for language, especially within broader semiosis and modelling, are necessary.

In order to offer an initial sense of the way that biosemiotics addresses the role that 'language' plays in relation to humans as a species, a long quote is illustrative. The following is from a footnote offered by Favareau and Kull (2015: 14):

P. Cobley, *Cultural Implications of Biosemiotics*, Biosemiotics 15,
DOI 10.1007/978-94-024-0858-4_3

The authors wish to note at the outset that use of the word language in this text should not be read to indicate a view of "language" as a reified "thing in itself", over and apart from the actual interactions between humans that give rise to such reified notions. The authors of this article find it important to emphasize the centrality in biosemiotic thinking that languaged behavior, as a subset of semiosic behavior, inherits all the latter's definitional properties, and that thus it could not be otherwise that sign behavior of all sorts is grounded in situated, actually instantiated action at all points – just as both Charles S. Peirce and Jakob von Uexküll each independently observed, approximately 100 years ago. If there is a view of "language" that holds it to be either a "thing in itself" or, indeed, anything other than semiosic behavior in its fully enacted cycles of perception and action (*Funktionskreis*) and unceasingly dynamic and emergent sign-object-interpretant (the last relatum of which must always be an action or an enacted change) interaction by and between living agents, it is not one held by us, or any other biosemioticians who are sufficiently conversant in the works of Sebeok, Uexkull and Peirce. Moreover, and as with the higher level term, semiosis, we feel that readers of the present volume will be sophisticated enough in such matters as to clearly understand that our use of the word language here is referring to an enacted and emergent semiosic process that arises solely from the interactions of living beings, and not with some misguided, fundamentally anti-biosemiotic notion of a "thing in itself". We include this note here for so as not to be mis-read, and to remind our readers to understand the following uses of the term language as shorthand for "linguistically-aided semiosis" – with all the ineliminable interactivity and triadicity that sign action in the world involves.

This is an important statement because even while "linguistically-aided semiosis" calls for a definition of the term 'linguistic' as much as 'language' calls for a definition of 'language', it makes clear that language is neither 'chatter', figures of speech or the like, nor is it the reified grammar witnessed in formal models, nor is it a cerebral substrate.

The idea of language as a reified grammar has roots that go back beyond Chomsky's *Aspects of the Theory of Syntax* (1965) through his review of Skinner (1959) all the way to Bloomfield (see Matthews 1993). The idea of language as focused on figures of speech or 'language games' has a more nebulous and demotic existence. Although this is not on its own cause for suspicion, inherent in the idea is an unrigorous formulation of 'language' as mere 'chatter'. By this is meant 'language' taken to consist of a series of tropes, figures of speech and vicissitudes of verbal communication. Such an approach to 'language' can be seen in middlebrow discussion in Sunday newspaper supplements and popular books on language (see, for example Steiner 1975; Burgess 1993; Ingram 1992). It is also evident in some linguistic specialisms, especially the focus on national languages rather than on the cognitive phenomenon of language. It is precisely the "use of the word *language* in which it is pluralisable in English" (Harris 1981: 12) that is at the root of what Harris identifies as the 'language myth' and which he argues has perpetuated disciplinary specialisms and a fixed code fallacy (see Chap. 6, below). Commonly, the important distinction between language as a cognitive capacity and the verbal interaction which is one of that capacity's manifestations is not made. So, any cognitive considerations in 'chatter' that remain knowingly or unknowingly embrace the Sapir-Whorf hypothesis or linguistic relativism too readily (for example, Bragg 2003; Bryson 1990; McCrum et al. 2002; Deutscher 2010). In such popular accounts, it is implicit that language as manifest in speech is 'special' and little heed is given to other forms of semiosis.

Favareau and Kull (2015) later go on to affirm that biosemiotics' main interest is actually in pre-linguistic semiotics. That is, biosemiosis focuses to a great extent on forms of semiosis that are more lowly, in terms of sophistication, than the human's facility for linguistic semiosis. Already, then, the problem of the relation between the linguistic and non-linguistic is raised. The history of sustained focus on these issues goes back to the immediate post-Second World War period in which a broad 'communication science' was being worked through by engineers, information theorists, cyberneticians, administrative researchers in media, mass communication theorists, political scientists, anthropologists, linguists and others. One of the key personnel in many of these meetings was the linguist, semiotician and biologist *manqué*, Thomas A. Sebeok (see, for example, Greenberg 1963; see also Cobley et al. 2011). However, the most concerted, focused and well-known addressing of the degree/kind question in linguistic terms was offered by Charles Hockett. His contention (1963: 1) that "A language universal is a feature or property shared by all languages, or by all language. The assertion of a (putative) language universal is a generalization about language" is problematic from a biosemiotic perspective because it elides the distinction of national and natural languages. Nevertheless, his (ultimately) 16 design features of language (Hockett 1963) -

> 2.1. Vocal-auditory Channel; 2.2. Broadcast Transmission and Directional Reception; 2.3. Rapid Fading; 2.4. Interchangeability; 2.5. Complete Feedback; 2.6. Specialization; 2.7. Semanticity; 2.8. Arbitrariness; 2.9. Discreteness; 2.10. Displacement; 2.11. Openness; 2.12. Tradition; 2.13. Duality (of Patterning); 2.14. Prevarication; 2.15. Reflexiveness; 2.16. Learnability (the first number for each design feature refers to the section of the essay)

- have been useful for many who have wanted to find points of distinction between human language and the communication systems of animals, plus the difference between what is a language and what is another form of communication (drum and whistle systems, monastic sign languages and the like) or merely a component of language.

In addition to identifying these features, Hockett was also able to make the following assertions:

3.1 Every human community has a language.
3.2 No species except our own has a language.
3.3 Every human communicative system usually called a (spoken) language is a language in our sense.
3.4 Every human language has the vocal- auditory channel (2.1)
3.5 Every human language has tradition (2.12).
3.6 Every human language has learnability (2.16).
3.7 Every human language has both an intonational system and a non-intonational system
3.8 In every human language, plerematic patterning and cenematic patterning are both (independently) hierarchical.
3.9 Human languages differ more widely in cenematics than in plerematics.
3.10 Human languages differ more widely, at least in their plerematic subsystems, at small size-levels than at large (again, the first number for each refers to the section of the essay).

Hockett makes some qualifications which are relevant to the present discussion in the current chapter. For, example he excludes writing from his consideration of language because it does not accord with 2.3, fading, and sign language is found not to be a language. Semiotics, on the other hand would most likely include writing in any definition of what is verbal and would point to the demonstration by William C. Stokoe, in the early 1960s after Hockett, of the grammatical properties that constitute ASL as a language (see Maher 1997). However, there are aspects of Hockett's observations that are of considerable importance for semiotics. Among these are Complete Feedback – the sender having access to the message; Semanticity – the power of denotation; Arbitrariness – conventionality in the service of reference; Discreteness – significance arising from the principle of difference; Displacement – signification in remote fashion of other times/places, fictions, ethics; Openness – the production of wholly new utterances, recursion; Tradition – an interesting one for biosemiotics, as will be seen, because it involves discussion of passing down language by 'natural' and 'cultural' means; Prevarication – potential falsity and meaninglessness of messages; Reflexiveness – signs about signs. No doubt other design features are important for semiotics, too, but these ones in particular stand out. Similarly, some of Hockett's broader assertions are borne out by biosemiotics while others are found questionable or less significant. "No species except our own has a language" is now the base point for discussions of language in evolutionary perspective, but "Every human language has the vocal-auditory channel" would be somewhat more open to question.

There have been criticisms of Hockett's system of design features and the applicability of one or other of them since they were first developed. Most recently, Wacewicz and Żywiczyński (2015) in the journal *Biosemiotics* have pronounced Hockett's design features to be a "non-starter". For them (2015: 42), his system is "radically unfit for capturing the difference between the communication of human and non-human animals from an evolutionary perspective". Hockett, they argue, focuses almost exclusively on the structure and medium of communicative signals rather than attending to the cognitive, social and ecological frame in which language use occurs. They call for greater consideration of the socio-cognitive and anatomical skillset stepping stones "that are not directly visible in communication but are the necessary prerequisites" (2015: 42). Effectively, Wacewicz and Żywiczyński call for a cognitive approach which, in contrast to a perspective which privileges patterning of messages, pays due attention to the processes of mind, sociality and sensorium which allow language to arise. As such, they are insisting on a contemporary semiotic line of investigation as opposed to a traditional linguistic one. In its broadest sense, this seems to be the crux of current criticism of the conception of language universals. Evans and Levinson (2009), for example, draw attention to the fact that language must exploit existing brain machinery and that the evolution of language is a matter of pre-existing developments of humans that are not necessarily specific to language in its current use. The tenor of these criticisms of language universals is an indication of the difference between two conceptions of semiotics – a common (but erroneous) one in which semiotics is taken to be devoted to analysis of the internal characteristics alone of texts, and a more contemporary conception of semiotics, fuelling the substance of these criticisms of universals, in which phenomena are approached with reference to their embedding in cognitive, social and material relations.

A recent essay by Deacon (2016) exemplifies this latter perspective. Although not directly addressing Hockett and universals, Deacon's essay nevertheless exposes some serious problems with regard to the emphasis of Hockett and many linguists on "plerematic patterning" and "cenematic patterning" in conceptualizing language. Treating language in terms of the formal system which seems to organize its manifestations is of a piece with approaching it with an engineering model, argues Deacon. He observes that language and language abilities have evolved spontaneously; because of this, identifying linguistic units as the building blocks of language is misplaced. Cognitive, semiotic and pragmatic structures generate linguistic units; elementary phonetic and morphological elements of language are "late-stage developments" (2016: 4–5). Deacon also makes this point with reference to brain development, demonstrating that cortical and language functions can only be understood in process terms, that the language function develops in a manner homologous to sensory and motor processing in general and, especially important, that the linguistic phrase is a semiotic, not linguistic, unit because it is lodged in a process involving the constraint of an index that enables symbolic reference. What is crucial here is not just the embedding of 'linguistic units' in semiotic processes but the fact that the 'linguistic unit' – which is always a broader semiotic unit in any case – could never make reference if it was not bound to an indexical operation. Put another way, it allows the possibility of a complex shifting from mind-dependence to mind-independence, in Deely's terms. Deacon concludes that there is a need for engineering logic to be reserved for the study of machines and organic logic applied to brains and language. "Paradoxically", he writes (2016: 24),

> the successes of formal generative linguistic theories may have impeded progress toward understanding language neurology and language evolution, even while they have provided remarkably sophisticated tools for the description of language structures. This is because the apparently remarkable adequacy of formal models to account for the complexities of language structure have contributed to an unwarranted assumption that language can be studied as though its structure was designed by a kind of instruction logic, as are other formal systems. Yet despite compelling evidence that language has a formal structure consisted with top-down rule-governed systems, its status as an evolved biological phenomenon raises serious questions about the plausibility of extrapolating from this descriptive analysis to a theory of language processing.

Deacon's bold and persuasive argument demonstrates that it is not only important to ensure that investigations of human semiosis are properly semiotically situated, but that they are also utilising the right tools. In this case, engineering tools will not do. One can see, also, without too indecorous a leap of the imagination, how Deacon's biosemiotic perspective has direct implications regarding how culture is interrogated. Evaluating human phenomena by means of a systematizing of their *manifestations* may not be productive when those phenomena ultimately have an organic base.

It is for these reasons that biosemiotics has made the process of modelling integral to its debates. The impossibility of enumerating and bringing into a closed system human processes of knowing, observing and experiencing suggests both that another conceptualization of them is needed and that language is similar in this respect. As has been seen, the conceptualization which biosemiotics has found is *Umwelt*, from von Uexküll, which fits the bill for a number of reasons. Firstly, it involves a theory of semiosis which applies to all animals. Secondly, it is organic and embodied, rooted in the sensorium/sign processing capacity of the animal. Thirdly, and connected to the second point, it involves physical manifestations yet the provenance of these in an organic source means that systematization of those manifestations alone is of limited explanatory value. Fourthly, it is realistic about the fallibility of species-specific semiosis; that is, while animals can never know the world in its entirety through their sign processing capacity, it is possible for them to know enough about it to effect their own survival.

The human *Umwelt*, according to Sebeok, drawing upon von Uexküll, is a model; or, put another way, various acts of modelling on the side of the *Innenwelt* (the inner, subjective world of the animal) contribute to the constitution of the "objective" or "public" world of an animal species as *Umwelt*. Models are made up of signs: thus, semiotic systems are modelling systems. Put simply, modelling is "how the organism (via its *Innenwelt*) maps the world, and what, for that organism, the meanings of the objects are within it" (Kull 2010: 43). The human *Umwelt* – like that of all other species' – derives from the sensorium, making semiosis a sensorial/cognitive activity. Modelling, in this formulation, whilst species-specific, is a criterial attribute of life. Thus, when the human *Umwelt* is found to amount to 'language', some qualification is needed. It has to be absolutely clear that 'language' in this formulation is not just 'chatter', nor could it ever be, and, in fact, that it is not even as closely related to 'chatter' as is commonly thought. Sebeok (2001a: 14) puts it like this:

> All known living organisms communicate exclusively by nonverbal means, with the sole exception of some members of the species *Homo sapiens*, who are capable of communicating, simultaneously or in turn, by both nonverbal and verbal means.
>
> The expression 'by verbal means' is equivalent to some such expression as 'by means of speech', or 'by means of script', or 'by means of a sign language' (e.g., for use in a deaf group), that are, each, manifestations of any prerequisite natural language with which human beings are singularly endowed. However, not all humans are literate or can even speak: infants normally do develop a capacity for speaking, but only gradually; some adults never acquire speech; and others lose speech as a result of some trauma (e.g., a stroke) or in consequence of aging. Such conditions notwithstanding, humans lacking a capacity to verbalize – speak, write, or sign – can, as a rule, continue to communicate nonverbally.
>
> […] The word 'language' is sometimes used in common parlance in an inappropriate way to designate a certain nonverbal communicative device. Such may be confusing in this context where, if at all, 'language' should be used only in a technical sense, in application to humans. Metaphorical uses such as 'body language', 'the language of flowers', 'the language of bees', 'ape language', or the like, are to be avoided.

Culminating in his 1988 essay on modelling, Sebeok's work attests that what characterizes humans is not the commonplace post-Chomskyan argument that they possess the capacity for language (again, sometimes simply reduced to speech or 'chatter' rather than the potential for recursion) but the possession of a verbal faculty *and* a non-verbal faculty. Put another way, in a compelling phrasing, humans are "apes *plus* language" (Deacon 1997: 5). Sebeok shows that what is known about early humans provides some important evidence for such a classification. It is thought that early hominids (*Homo habilis*, about 2 million years ago) harboured a' language', grammar or modelling 'device'in their brains. *Homo erectus* (about 1.5 million years ago), with an increased brain size over his/her predecessor, also possessed the capacity, an as yet unrealized ability to develop a sophisticated human verbal communication system. However, verbal encoding and decoding abilities only came into use about 300,000 years ago with early *Homo sapiens*. Two conclusions arise from this. Firstly, if language appeared in humans so early, then it provides grounds for the idea that it is was involved in a long period of co-evolution with the brain – "Languages also have to adapt brains" – as posited by Deacon (1997, 2012b: 33). Secondly, it indicates that humans therefore possessed the capacity for *language* long before they started to implement it through *speech* for the purposes of verbal *communication*. Prior to the verbal form, communication would have taken place by nonverbal means, a means that humans continue to use and refine today (see Sebeok 1986b, 1988). *Homo habilis* and *Homo erectus* therefore appear to have had what Sebeok, following his 1988 adjustment of Lotman's formulations, calls 'primary modelling'. *Homo sapiens sapiens* evolved secondary and (as the inevitable consequence) tertiary modelling.

The primary modelling system is the key concern, here, although the tripartition of secondary and tertiary modelling do need to be considered. Primary modelling in humans is the capacity for verbal and non-verbal communication. Secondary modelling is driven by speech – that is verbal *communication through vocal interaction*; this is one aspect of verbality in general which includes written as well as spoken signs. *Pace* both Hockett and Evans and Levinson (2009) who state that language requires a vocal tract, semiotics sees secondary modelling in evolution as the result of an exaptation (Gould and Vrba 1982), not an adaptation, of the primary modelling capacity for specific interactions among humans. That is to say, speech was not an inevitable developmental consequence of the language capacity; instead, it was co-opted, presumably at the moment that environmental conditions made it most expedient.

Tertiary modelling, then, is the extension, through inevitable mutation in social exchange, of primary and secondary modelling to produce cultural forms (including predominantly verbal ones – e.g. novels; non-verbal forms – e.g. paintings; mixed forms – e.g. theatre) which not only partake of the lower strata of modelling but also feed back to them (see Sebeok and Danesi 2000). Put another way, tertiary modelling has secondary and primary modelling nested in it. The point to be made about primary modelling – a point that runs through Sebeok's post-1979 work but which was never offered in a definitive formulation – is that it is not only *not* a matter of 'language' conceived as 'chatter', but that it is not even a matter of communication conceived as message transfer (as might be discerned in the non-verbal communication of hominids). Instead, primary modelling is an acute and developing cognitive capacity to differentiate within an *Umwelt*. The more humans differentiate, the more they enhance their *Umwelt*. Sebeok (1979a) identified the first stirrings of differentiation or modelling at a threshold as lowly as the cell in its interactions, by way of the immune system and anxiety, with bodies outside itself (see Chap. 4, below). Far more complex forms of differentiation of an *Umwelt* are apparent in the aesthetic behaviour of animals (Sebeok 1979b; see Chap. 8, below). That differentiation is a (by-)product of proliferating semioses which inevitably incur sociality. Since taking signs in isolation is invariably an act of extreme abstraction, the object of semiotics, even biosemiotics is concerned with this sociality of semiosis (see Cobley and Randviir 2009). In a fashion that is actually allied to the generative aspect of the Chomskyan perspective (cf. Augustyn 2009), Sebeok argues the evolutionary advantage of increasing differentiation or distinction of features of an *Umwelt*. The ability to perceive relations (as distinct from related things) and to cognize a superior number of elements of the world is seen as characteristic of humans – this is where language defines what it is to be human; and this is where sociality – the interconnectedness of signs that humans are able to apprehend – is crucial to the process.

In Sebeok's conception of modelling, sociality's role is implicit. In some way, human sociality seems to have stimulated hominids to differentiate the natural world ever more extensively. 'Language' is thus special, rooted in the doings and development of hominids. However, it is not to be considered as removed like a sovereign from all animals' modelling. It contains elementary modelling processes as part of the more sophisticated process. In their mapping of the three forms of modelling onto Peircean categories, Sebeok and Danesi (2000) associate primary modelling with Firstness, possibility and abduction. As such, its powers of differentiation that make up the human *Umwelt* are built first in the realm of quality. Primary modelling is a fitting home for qualia both ontogenetically and phylogenetically, a 'pre-social' mode of existence for young humans and young humanity alike: "The ability to have sense experiences and to be able to distinguish between qualitatively different ones (qualia) – sweet and sour, hot and cold, red and green – is basic to knowledge, understanding, communication, and intelligent reasoning" (Brier 2008a: 38). These 'basic' sense experiences are of a piece, or continuous, with the 'higher' mental processes. Primary modelling is the sphere of affect, of qualia on the way to differentiation and the well-spring of motivation which is moulded, from its combined verbality and non-verbality, in secondary modelling. In the process, parts of primary modelling are routinely forgotten by humans (see Chap. 7, below). Nevertheless, understanding secondary modelling surely calls for an examination of its provenance in primary modelling rather than a formal analysis alone of secondary modelling's units.

In addition to its development in sociality, primary modelling is arguably the crucible of 'first-person experience', where feeling develops in tandem with recognition, memory, categorization, mimicry, learning and communication. First-person experience, it is worth remembering, is fallible. It can be characterized by misrecognition, amnesia and miscommunication. For Sebeok, it is axiomatic that "what a semiotic model depicts is not 'reality' as such, but nature as unveiled by man's method of questioning" (1991a: 12) that is, the tangled relations of *ens reale* (mind-independent being) and *ens rationis* (mind-dependent being) in semiotically explicated scholastic realism (see Chaps. 1 and 2, above and Cobley 2009). He adds (Sebeok 1991a: 17–18),

> In the age-old philosophical quest for reality, two alternative points of departure have been suggested: that the structure of being is reflected in semiotic structures, which thus constitutes models, or maps, of reality; or that the reverse is the case, viz., that semiotic structures are independent variables so that reality becomes the dependent variable. Although both views are beset by many difficulties, a version of the second view, proposed by the remarkably seminal German biologist, Jakob von Uexküll, under the watchword *Umwelt-Forschung* – approximately translated as 'research in subjective universes' – has proved to be in best conformity with modern semiotics (as well as ethology). The same attitude was expressed by Niels Bohr when he answered an objection that reality is more fundamental than the language which it underlies. Bohr replied: 'We are suspended in such a way that we cannot say what is up and what is down' (French and Kennedy 1985: 302). Signs have acquired their effectiveness through evolutionary adaptation to the vagaries of the sign-wielder's *Umwelt*.

Furthermore,

> A complicating fact of life is that the bare act of observation entails a residual juncture that disturbs the system being observed. The essential ingredient, or nutriment, of mind may well be information, but to acquire information about anything requires, via a long and complex chain of steps, the transmission of signs from the object of interest to the observer's central nervous system. Its attainment, moreover, takes place in such a manner that this influential action reacts back upon the object being observed so as to perturb its condition. In brief, the brain, or mind, which is itself a system of signs, is linked to the putative world of objects not simply by perceptual selection, but by such a far-off remove from physical in-puts – sensible stimuli – that we can safely assert that the only cognisance any animal can possess, 'through a glass darkly', as it were, is of signs.

Sebeok, as ever, was concerned with semiotics as a matter of negotiating illusion and reality, a matter of how to know which is which and how they are related. Much of human modelling is situated in such a way that it is remote from the physical realities of mind-independent entities. This has its advantageous features in that it enables humans, as part of their remote processing of semiosis, to imagine possible worlds, fictions and ethical scenarios. Of course, if humans could *only* model in a remote fashion, completely unconnected with mind-independent physical reality, they would run the serious risk of extinction.

As a definition of language, language as modelling is simply more ambitious than erstwhile pioneers in language as communication (e.g. Bühler 1990), language as differentiality (e.g. de Saussure 1983), language as discourse (e.g. Vološinov 1990), language as sociality (e.g. Halliday 1978), language as communication of the laity (e.g. Harris 1999) or language as indigenous to humans (Hockett 1963) even while agreeing with parts of those theses. Its nearest relative is, as mentioned earlier, the generative aspect of early positing by Chomsky of universal grammar. Language as modelling differs from all of these other perspectives in that it explicitly entails an evolutionary theory of what language is, incorporating a fully semiotic account of its sociality and cognitive relations. Therefore, it is not solely reliant on the analysis of plerematic patterning and cenematic patterning. In light of Sebeok's thought, this is to be expected. In addition to being a biologist *manqué*, Sebeok proposed that semiotics is a systems theory (see Sebeok 1977), dedicated to unravelling semiosis across all realms, with a project of de-ontologization. The systems theory aspect of Sebeok's semiotics latterly made references to physicist John Archibald Wheeler's notion of the participatory universe. Indeed, Wheeler is a recurring presence in Sebeok's later books (1986a, 1991b, 2001b) and his "highly imaginative rendition of the so-called Copenhagen interpretation" (2001b: 38) is referenced to demonstrate that the universe "viewed as an autopoietic 'self-excited circuit' is necessarily dependent on life, 'mind', and observership" (2001b: 16). What matters to Wheeler, and is taken up by Sebeok, is not so much the anthropocentrism of observation so much as the scale involved and the measuring device in observation. This is summed up in the Sebeokian title of Wheeler's famous essay: 'Not consciousness but the distinction between the probe and the probed as central to the elemental quantum act of observation'. There (Wheeler with Ford 1998: 330), he suggests that

> No matter what the uncertainties of the small-scale world, no matter how chaotic the fluctuations, our knowledge of nature rests ultimately on perfectly definite, unambiguous observations – what we see directly or what our measuring apparatus tells us. How can this be? If the world 'out there' is writhing like a barrel of eels, why do we detect a barrel of concrete when we look? To put the question differently, where is the boundary between the random uncertainty of the quantum world, where particles spring into and out of existence, and the orderly certainty of the classical world, where we live, see and measure?

This quote is of a piece with the concept of *Umwelt* as a de-ontologization. *Umwelt* is a matter of species 'knowing' in the process of modelling: a human may know something as a barrel of concrete. But *Umwelt* is also a matter of *being* for the simple reason that an it is a species' very mode of inhabiting: it is possible to experience something as eels in, say, terms of their harmful or beneficial attributes even while detecting them as a barrel of concrete. Certain kinds of knowing or observing, through signs, can interact with being to deliver humans from mind-dependent reality to mind-independent reality.

In addition to Sebeok's co-opting of Wheeler, biosemiotics also considers the role of the human as observer in the idea of the 'semiosphere', a concept which complements modelling. Derived from Lotman's cybernetic and autopoietic formulation, its definition is extended by Kull when he suggests that "Semiosphere is the set of all interconnected Umwelts. Any two Umwelts, when communicating, are a part of the same semiosphere" (Kull 1998: 301). Thus, a domestic cat and its owner share the same semiosphere when they are each eating a portion of a fish that the latter has cooked for both of them. For feline and human, the fish is a component of what they understand as food. However, the ways that these two members of different species will relate to the food, how the food exists in their *Umwelten*, are very different – the cat's eating may be solely for survival, it may be totally dependent on its owner; the human might eat simply for pleasure, for specific gustatory experience, to partake of a cultural and culinary pursuit, to exercise some knowledge of the history of the fish and members of its species. It follows that the cat as observer and the human as observer are obviously very different. The main difference is that the human *knows* that s/he knows. The human is capable of understanding of Thirdness relations *qua* Thirdness relations (Favareau and Kull 2015: 16 n. 11). To put it another way, the human animal is a semiotic animal (Deely et al. 2005; most systematically in Deely 2010) defined by the fact that s/he knows that there are signs.

Human knowing of the existence of signs, however, does not, on its own, necessarily entail a theory of the observer. There is still a need to develop the latter, as Brier (2008a: 119) insists:

> There are many arguments – even outside cybernetics and system science – for this philosophy of science and its basic epistemological conceptions to begin with the observer or the phenomenological position, thereby acknowledging that humans are knowers, even if we do not know why or how. It is important to acknowledge our existing ignorance in the area of what it is to know and how knowledge comes about. We must also acknowledge that we are observers co-existing in language with other humans in culture and society.

Cybersemiotics, as a key part of biosemiotics devoted to questions of knowing (see Chap. 1, above), has endeavoured to institute this perspective and to move beyond the second-order cybernetic formulation that the result of human observation of the universe from the vantage point of its *Umwelt* is somehow 'undecidable'. The noetic perspective on the realism dilemma introduced by semiotics suggests that the world in given particulars at least will finally be decidable, and that semiosis in *Umwelten* (at least in the *human Umwelt*) is workably reliable until such a time when greater efficiency in apprehending reality might be attained. In contrast to the nominalism of, say, poststructuralism, which figures humans as dominated by their sign systems, Sebeok's work, for example, is a prolonged realist but not naïve consideration of the spiralling complexity of human semioses and the understanding of them coupled with a rueful awareness of how the institutionalization of knowledge and phylogenetic forgetfulness impede that understanding (see also Cobley 2011).

Biosemiotics in the wake of Sebeok is therefore explicitly future-orientated. It is constituted, as John Deely has explained (most recently, 2015b) by a *vis a prospecto*, taking its cue from Peirce's focus on science as the possibility of projection and the predictive capability of laws, along with his theory of the potential engendered by the interpretant in anticipating future developments and considering even those potentialities that remain hidden. The indications of Sebeok regarding the participatory universe reveal that biosemiotics also has a nascent theory of the observer. Kull's (2009) short contribution regarding Σ-sciences and Φ-sciences, the latter focused on universal laws and quantitative methods, the former concerned with local semioses and using qualitative research to investigate how organisms 'know', is a preliminary signpost in the return to Sebeok's concerns in this respect. Biosemiotics is placed firmly on the side of the former, a science of knowing rather than a science of laws. Along with this, there is the need to conceptualize the circumstances of 'knowing' in a rigorous fashion, for the 'knowing' of species does not occur in a vacuum. Central to biosemiotic theorizing has been the goal of non-anthropomorphic observation – investigating on the terms of the species in question rather than in human terms. From Sebeok's discussions of the Clever Hans phenomenon in particular (Sebeok and Rosenthal 1981), there has been vigilance in semiotics regarding anthropomorphism and the pathetic fallacy. In the past, avoidance of anthropomorphism was often like a plea of 'undecidability', that it was impossible to impute motivation or purpose in the constitution of an animal or plant (as if it was possible to do so with a human animal) and what remained was to simply catalogue the organism's interactions with its environment. Recently, however, both teleology and purpose have been reconsidered, especially in biosemiotics (Deacon et al. 2009). In relation to this, it seems that both the *Umwelt* and the semiosphere contribute to fulfilment of the biosemiotic concept of the 'knowing' environment.

In biosemiotics, arguably, difference is a key factor. The idea of 'semiotic freedom' (Hoffmeyer 1996), the ideas of agency (Hoffmeyer and Kull 2003), species specificity (Uexküll 1992) and Sebeok's 'differentiation' in relation to primary modelling, all emphasize the differences of semiosis and the feedback on semiosis of species. *Difference* is the pivot around which observation in biosemiotics could be further theorized. Certainly, it is difference that is in play in relation to the objects of biosemiotics, and it should be stressed at the same time that in the theory of *Umwelt* the object is never neutral for species (if not '+' – to be sought out – or '–' – to be avoided – the object is '0' – safe to ignore or of no immediate interest). The difference that is entailed by the non-neutrality of objects in recognition, memory, categorization, mimicry, learning and communication points to a dimension of knowing or "environment", but not just in the sense of the physical environment of organisms. The relations between the sensorium of the organism and whatever impinges on the sensorium is crucial. Deely (2001: 127–8) explains:

> Uexküll uniquely saw that the difference between objects of experience and elements of sensation is determined primarily not by anything in the physical environment as such but by the *relation* or, rather, network and set of relations that obtains between whatever may be *in fact* present physically in the surroundings and the cognitive constitution of the biological organism interacting with those surroundings here and now. Nor are those relations primarily of the type that antecede and hold independently of any such interaction. To the contrary. The relations in question are not mainly between the organism and what is sensed, those limited and partial aspects of the physical surroundings which are proportioned to and activative of the limited range of this or that sensory channel in combination with however many other cognitive channels the organism in question is biologically endowed with. No. The relations in question concern above all how the limited and partial sensory aspects of the physical environment are connected among themselves so as to constitute *objects of experience*, and this constitution depends above all on the constitution of the organism doing the sensing. For it is the interests of that organism, not the 'independent' nature of the source of the sensory stimuli, that is at issue in the perception as such that the organism finally acts upon and uses to orientate itself within the environment for the purposes of its life and well-being.

The distinctions made here are important: as was discussed in Chap. 2, above, an animal's *Umwelt* is its 'objective' world – not 'subjective' as is so often assumed – and it is where an animal relates to 'objects'. The reason for this, fittingly, pertains to Deely's logical re-figuration of objectivity through the rigorous formulation of 'thing', 'object' and 'sign'. He demonstrates that the world that seems to be wholly independent of humans – in the common sense of 'objective' – can never be such. Rather, it is a specific kind of mixture of that which is independent of, and dependent on, humans (or fluctuating between mind-independence and mind-dependence).

Thus, any object, including the objects which are the focus of science, is dosed by experience. As such, science does not really deal with 'thinghood' pure and simple, no matter how extensively physicalist science purports to be concerned with the ontology of 'things'. In general, the concern of scientific research is objects, objects presenting themselves as things or, if the science in question is progressive and self-aware – that is, taking account of the observer, its own process of observation – objects on the way to revealing things.

In biosemiotics, Peirce provides an evolutionary philosophy, a cosmology and a sustained consideration of purpose, causation and finality (see, for example, Hoffmeyer 2008a, b). All of these are germane to the consideration of the observer but they also serve the task of banishing anthropomorphism in pursuit of *Σ*-sciences and a theory of knowing which will enable the apprehension of such phenomena as agency in nature (see Chap. 4, below). The banishing of anthropomorphism in the *Φ*-sciences (such as behaviourism), however, and in aspiring *Σ*-sciences (such as biosemiotics) are two different matters. In its commitment to a science of knowing, its pursuit of 'meaning' *for* species (no mere sop to Cerberus) and its acknowledgment of the impediments that anthropomorphism has created, biosemiotics looks in the opposite direction from the Laplacean, mechanist, determinist overtones of physicalist science and the instrumentalist parts of the research establishment which uphold it. In its undergirding by the theory of *Umwelt* and semiotics, too, biosemiotics has, as argued already, a theory of the observer which cybersemiotics has also attempted to develop. So, in respect of the human, there is a difference from other organisms since the human is aware of the difficulties insinuating themselves into observership whereas other organisms do not consider the practice of observing. While humans share with other organisms a process of 'knowing', humans can imagine beyond their sphere of immediate observation of the material phenomena they encounter. Indeed, the anthropomorphism of human observation may even be a beneficial part of scientific work in that it could require examination of the relations of conflict in the natural world, initiated or engendered from either direction, between humans and other organisms. Alternatively, it might involve the re-visiting of evolutionary principles in the theory of *Umwelt*, for while the survival of species is an indication of the workability of their modelling system, "Neither survival nor maximal dissipation of entropy is enough to explain the growth of systems with inner worlds of qualia" (Brier 2008a: 377). Certainly, it would involve attention to those *relations* which obtain between the 'knowing' accruing to organisms and the knowing by which they might finally be known.

In reference of human modelling, language and the difference of humans from other species, biosemiotics is neither reductively idealist in the modern sense nor *naively* realist in a premodern or modern sense. What is at issue in biosemiotics is not human 'perception' set against 'the real', but the *relations* obtaining between reality and illusion, between brute physicality and the sensoria of beings. These are the relations that are in play in the ongoing process of modelling, a component of which is the possibility of knowing. If all organisms can be investigated in terms of their possibility of 'knowing', then humans are only different in degree; but, of course, in its very commitment to investigating according to this difference in (semiotic) degree, biosemiotics reveals that humans are also different in kind. Yet it should be emphasized that the difference in kind of human observership is not simply a matter of knowing that there are signs and that reality might be 'undecideable', as in the the tired resignation of, say, constructivism. As Eco puts it (1990: 41),

> if the sign does not reveal the thing itself, the process of semiosis produces in the long run a socially shared notion of the thing that the community is engaged to take as if it were in itself true. The transcendental meaning is not at the origins of the process but must be postulated as a possible and transitory end of every process.

It is not even the social sharing that is important here so much as the fact that semiotics looks forward, with a *vis a prospecto*, envisaging what work can be done. It anticipates that the observer resides in a (Peircean) community of interpreters. This might seem to coincide with von Foerster's constructivist conclusion that "Reality = Community" (2003: 227). However, in biosemiotics, the matter is less simple than this formulation, even with its implications, allows. Von Foerster (2003: 216) argues for the use of the indefinite article in the noun-phrase "a reality", distinguishing between the two uses of the article thus:

> The 'The-School': My sensation of touch is *confirmation* for my visual sensation that there is a table. The 'A-School': My sensation of touch in *correlation* with my visual sensation generate an experience which I may describe by 'here is a table'.

Apart from the fact that he merely supplements one sense with another in retreating into experience from the possibility of the real, von Foerster has set up another false opposition as far as semiotics is concerned. Semiotics is of the 'the' school, but not because it relies on sensation as confirmation but because it is concerned with the *relations* between sensoria and brute physical reality. What biosemiotics indicates as characteristic human modelling is its inherent equilibrium of experience such that the observer will be able to get as close as makes no difference to the thing which is beyond the sign and the object, the thing which is at the boundary of the *Umwelt.*

If it is considered that this is what science is trying to do, then religion, art, fiction, politics and ideology, by comparison, seem relatively impoverished ways of knowing. They are first-person experiences, quasi-knowings, which do not evince a meaningful impulse to move beyond signs or objects or, indeed, in some cases, beyond first-person experience. They seem to be the result of structures which are solely human in origin, a position which has been exploited by the ultra- or postmodern idea that the world is constituted only by shifts in power and that, effectively, nothing exists beyond the polis. Biosemiotics presents a synthesis of human knowing in which it is continuous with semiotic processes in other organism. That is not to say that the concept of 'knowing' fails to reveal anything beyond the fact that there *is* knowing or to challenge the idea that whereof one cannot speak, thereof one must be silent. Instead, it provides a crucial frame for explaining human modelling. In particular, and in consonance with Deacon (2016; see above), it prompts investigation of language as an organic phenomenon rather than an engineering process. It also makes necessary an appreciation of the broader semiotic – cognitive and social – constraints and affordances that make language 'special' and much more than the sum of its cenematic and plerematic components. For it is clear in biosemiotics that language is not just a result of straightforward determination in nature. Hoffmeyer notes how some of his students have been reluctant to accept that human language is something special within nature in general; he responds "While the most intimate, the most profound of human experience can, at all events, hardly be said to be of a linguistic nature, to maintain that language is nothing special is quite absurd" (Hoffmeyer 1996: 97). This observation does not amount to exceptionalism. However, the impasse of the engineering approach and the descriptive approach to language makes almost understandable the resort of some to an exceptionalist explanation of language and humans. Exceptionalist formulations regarding culture are at the very least metonymically linked to the same conception of language and humans. Biosemiotics' theory of modelling offers both an evolutionary perspective, featuring gradations and phases of development, along with the recognition that humans' symbolic reference, while not arising from thin air, is unique in the biosphere.

The common question arising from Darwin's original assertion has been whether humans and non-human organisms are separated by a difference in kind or difference of degree. As has been seen, without any fence-sitting – indeed, with a radical purpose – biosemiotics insists that humans are separated from other organisms by a difference in kind *and* a difference in degree. Hockett was forthright in his contention that while languages have a 'tradition' that is passed down the generations, this could never involve the conventions of a language being transmitted "through the germ plasm" (1963: 9). There is some bet-hedging here, probably resulting, too, from elision of the distinction of national and natural languages and his use of the indefinite article. Yet, even Hockett suggests of language that "Genes supply potentiality and probably a generalized drive, since non-human animals cannot learn a human language and humans can hardly be prevented from acquiring one" (1963: 9). So, while Hockett separates language from other animal communication, such as that of the bee which is most likely genetically inherited, there is consanguinity in the very process of genetic inheritance. Biosemiotics, particularly on the question of language, sees the genetic inheritance not as a matter of periodic mutations but as a development involving nesting and embedding. Language is characterized by a nesting or embedding of symbolic reference (Deacon 1997; see Chap. 1, above). That co-evolution of language and brain of which nesting and recursion (the "combinatorial possibility that comes for free, so to speak, as soon as symbolic reference is available", Deacon 2012b: 19) is a part, produces a capacity that no other animals possess. This 'difference in kind' offers the possibility of projection, a characteristic of the human *Umwelt* which should not be considered as a unit but as imbricated with the specific forms of recognition, memory, categorization, mimicry, learning and communication, as well as agency and scaffolding. It is the potential to produce both artistic forms, fictions (see Chap. 8, below) and ethics (see Chap. 5, below), to recognize, as the semiotic animal, Thirdness relations – that there are such things as signs. Indeed, because of the difference in kind of language, Deacon (2012b: 34) suggests that humans have created a "symbolic niche" for themselves in which there is flexibility and relative freedom from constraint, as well as, possibly, imagination, the arts and social conformity. Yet, it should be remembered that language, here, is primary modelling, a cognitive capacity for nuanced differentiation which has the potential for outward manifestations in verbal and non-verbal communication. Glottocentrism tends to forget or repress this dual heritage of human modelling even while culture repeatedly acknowledges it through its proliferation of mixed forms. Rather than attempting to discursify nature, living or dead, in the hope of producing any enhancement to human understanding of it, biosemiotics' implication for culture in this respect is the opposite of glottocentrism. The human phenomenon of language is shown to be special and 'different in kind' *not* for reasons of its departure from all other semiosis but precisely because of its qualified continuity with living nature.

Chapter 4
The Natural Subject

As well as its insights into what it is to be human, biosemiotics also re-formulates what it is to be a human subject. It upsets notions regarding the distinction between collectivity and the individual that have contributed to common sense in the modern world and especially since the French Revolution (see Siedentop 2015). Arguments regarding the human subject have been part of the burgeoning literature of 'identity studies' in the last 25 years. In the modern literature in this area, there is often a tension between what is referred to as 'selfhood' and what is understood as 'subjectivity' (cf. Atkins 2005: 1–2). The former, broadly, involves a conception of the human as *conscious* of its own existence and most of its intellectual capacities as well as its distinction from others; the latter, generally, has come to mean the human as *constituted* by the range of 'practices' which precede its existence and subsequently – or 'always already' – shape its existence, thought processes and options. Such practices are 'cultural' in their bearing or, more pointedly, signifying practices. What has probably become axiomatic in much of the writing on identity, the subject and the self in modernity is that subjectivity and selfhood are synonymous mainly because they are no longer considered to be unitary or intrinsically constituted in character (see Benhabib 1992; Cascardi 1992; Giddens 1991; Taylor 1992). Elliot writes (2001: 2):

> Selfhood is flexible, fractured, fragmented, decentred and brittle: such a conception of individual identity is probably the central outlook in current social and political thought. As the pace, intensity and complexity of contemporary culture accelerates, so too the self becomes increasingly dispersed. Displaced and dislocated within the wider frame of postindustrial capitalism, the individual self turns increasingly to consumption, leisure and travel in order to give substance to everyday life. Or so some have forcefully argued.

The coda here, of course, is crucial. To be sure, there are other ways to conceive of subjectivity and selfhood that embrace the concepts of choice, agency and flexibility whilst retaining an anti-humanist perspective on the limits to human freedom.

P. Cobley, *Cultural Implications of Biosemiotics*, Biosemiotics 15,
DOI 10.1007/978-94-024-0858-4_4

Furthermore, and as Elliott suggests, perspectives from "Freudianism and feminism to poststructuralism and postmodernism" (2001: 11) have saturated the study of subjectivity in modernity. Clearly, the tracing of such arguments from Descartes through the Enlightenment to feminism is not only Eurocentric but also privileges the categories of so-called 'continental philosophy'. Yet, within those perspectives has been the implementing of the work of Saussure, leading to a body of speculations on 'language' which has not only misunderstood or distorted its source (for a comprehensive overview, see Harris 2003) but has also been cripplingly limited in its purview. Some of the antinomies in definitions of language that were encountered in considering the difference in degree or kind of humans from animals recur in considerations of the human subject.

As with the continental philosophy tradition, the Cartesian perspective on subjectivity, which focused specifically on the *cogito* or the 'I', providing sustenance for individualism, has been found to be particularly incompatible with biosemiotics. As might be guessed from previous chapters in this volume with their description of biosemiotics' emphasis on continuity across nature, the Cartesian separation of mind and body functions is diametrically opposed to biosemiotics. This is not just because of the ways in which Cartesian theories, particularly in their application to culture, have given free rein to mental life while placing the body in a subordinate role. It is also because the subordination of bodily components has been consonant with the belief in the limits of these components. That is to say, Cartesianism is thoroughly invalidated by biosemiotic recognition of endosemiosis. Hoffmeyer notes that "the human being is really the product of a collaboration between possibly hundreds of trillions of bacteria!" (Hoffmeyer 1996: 23), seeing fit to put an exclamation mark on what was previously unspoken. He adds, "However strange it may sound, it seems likely that the modern-day eukaryotic cell was generated by some kind of symbiosis, whereby a great many tiny prokaryotic cells combined to form one large departmentalized cell, the eukaryotic cell" (1996: 30). Thus, to begin with, the feeling of unity so beloved of Cartesianism is an illusion (Hoffmeyer 1996: 86). The "One quadrillion bacteria, in the form of ten trillion cells, [that] collaborate on the job of being human" (1996: 124) constitute a 'swarm' (Hoffmeyer 1995) of such intensity that they "cannot live up to the demand placed on them to be meaningless" (1996: 128). Thinking beyond semiosis, but in continuity, Hoffmeyer later contends that humans are always subject to collectivization of *Innenwelten* through the speech facilitated by language, their *Umwelten* (1996: 112). The main point for the moment, though, is that endosemiosis demonstrates that the body is certainly not under the tutelage of the mind and, further, that, in a fashion that makes individualism untenable as an idea, the body is not 'our own'. In this way, biosemiotics has coincided with work on the *embodiment* of cognitive processes, the 'naturalization' of intelligence even as it explored the 'intelligence' of nature (Hoffmeyer and Kull 2003: 260–1).

While the human is the product of, or constituted by, endosemiosis, there is a need to be a little more specific, even in a general account such as this, regarding what it is to which the human *is subject*. This raises, too, what has also been seen to be the vexed issue of agency. For glottocentric approaches, it is clear that the human is subject to discourse or 'language' or 'languaging' or the 'play of the signifier' and so forth (see, for example, Belsey 2005). Yet, biosemiotics' understanding of language lies some distance from that of glottocentrism, in the conception of language as a modelling device laid out in the last chapter. That is, a device not necessarily connected to speech, innate in humans in a broadly post-Chomskyan fashion and constituting an *Umwelt* characterized by a potential for differentiation far in excess of other animals. In this sense, then, humans are subject to language. Crucially, though, humans as constituted by endosemiosis are subject to collectivity while they are also subject to an 'other'.

The first of these is visible in biosemiotics as a result of the insistence on synechism (see Chap. 2). Esposito (n.d.) notes that

> certain objects may be regarded as logical individuals or subjects of linguistic predication but when we are being natural philosophers we cannot afford to regard them as such because we run the risk of failing to recognize and ask important questions about them. Instead, they should be regarded as inexhaustible collections of systems with no a priori boundaries.

Peirce, likewise, invokes synechism in order to abandon the illusory and reifying effects of a priori boundaries. In a general, philosophical sense, he writes (6.173),

> It is not, upon synechist principles, a question to be asked, whether the three angles of a triangle amount precisely to two right angles, but only whether the sum is greater or less. So the synechist will not believe that some things are conscious and some unconscious, unless by consciousness be meant a certain grade of feeling. He will rather ask what are the circumstances which raise this grade; nor will he consider that a chemical formula for protoplasm would be a sufficient answer. In short, synechism amounts to the principle that inexplicabilities are not to be considered as possible explanations; that whatever is supposed to be ultimate is supposed to be inexplicable; that continuity is the absence of ultimate parts in that which is divisible; and that the form under which alone anything can be understood is the form of generality, which is the same thing as continuity.

More pointedly, he addresses synechism in relation to the human subject (7.571)

> Nor must any synechist say, "I am altogether myself, and not at all you." If you embrace synechism, you must abjure this metaphysics of wickedness. In the first place, your neighbors are, in a measure, yourself, and in far greater measure than, without deep studies in psychology, you would believe. Really, the selfhood you like to attribute to yourself is, for the most part, the vulgarest delusion of vanity. In the second place, all men who resemble you and are in analogous circumstances are, in a measure, yourself, though not quite in the same way in which your neighbors are you.

Biosemiotics, as has been seen, does embrace synechism and it does abjure this metaphysics of wickedness. Yet, synechism is not just a matter of philosophical method in terms of its outcomes, either for Peirce or biosemiotics. Tellingly, Peirce goes on directly to refer to "the barbaric conception of personal identity" (*EP* 2: 3). The strength of Peirce's denunciation in this passage indicates the pervasive nature of both philosophically standard accounts of selfhood and the common, demotic understanding of the self that makes up quotidian lived relations and which he felt compelled to oppose in no uncertain terms (Cobley 2014). Biosemiotics' insistence on continuity from endosemiosis and across nature to the sphere of human affairs is already political in its vivid counterposition to modern liberal humanism.

The second instance sees the human as subject to an 'other', a relationship which has a deep heritage. Sebeok identified the first stirrings of the self in relation to the other at a threshold as lowly as the cell in its interactions, by way of the immune system and anxiety, with bodies outside itself. Sebeok's notion of the 'semiotic self' provides a programmatic re-visioning of the concept of subjectivity, placing an emphasis on semiosis as characteristic of life. His exploration takes place mainly in four published essays beginning with 'The semiotic self' (1979a) and 'The semiotic self revisited' (1989), both reprinted as a brace in his 1991c book, *A Sign is Just a Sign*, and "Tell me where is fancy bred?': The biosemiotic self' (1992) and 'The cognitive self and the virtual self' (1998), reprinted as a second brace in the 2001 volume, *Global Semiotics* (2001b).

For Sebeok, not only is selfhood inaugurated at a lowly level of biological development, it also involves a very particular kind of sign. He begins his stimulating but laconic essay on 'The semiotic self' by pointing out the economy of Freud's definition of anxiety. For Freud and Sebeok, anxiety is a sign and, more specifically, a resolutely indexical one. It is integral to the workings of the immune system of an organism which maintains a distinction between 'self' and 'non-self'. According to Sebeok, the immune system harbours a kind of 'memory' based on biological discrimination, but also operates another kind of memory, anxiety, whose domain is patterns of behaviour. Anxiety is activated when the self is menaced and this can be triggered by signs that might "take a quasi-biological shape, such as the olfactory trace of a leopard predator for a baboon prey, or be of semantic character, such as some verbal assault whereby a stranger presses in upon the territories of the Self" (1991c: 39). Indeed, at the level of higher organisms, one precipitator of anxiety mentioned by Sebeok is the constraining factor by which physical symptoms are resistant to verbalisation and narrativisation in doctor-patient interactions (2001b: 123). The brute physicality or indexicality of symptoms demands interpretation, an activity which Sebeok also sees as central to the maintenance of self, inevitably in relation to an other:

> Any self can and must interpret the observed behavior of another organism solely as a response to *its* interpretations of *its* universe, 'behavior' meaning the propensity that enables it to link up its Umwelt with those of living systems within its niche (2001b: 126).

The act of interpretation, he adds, is an act of "as*sign*ment", that is, the elevation of an interpreted phenomenon to 'signhood'. It is also self-maintaining or 'autopoietic' (2001b: 126).

In his essays, Sebeok focuses on anxiety, love and the self-apprehension of body size in the maintenance of the self. Yet anxiety's role in the immunological system appears to be pivotal. It is worth replaying Sebeok's summarized propositions, here:

1. There are at least two apprehensions of the Self:

   (a) *Immunologic*, or biochemical, with semiotic overtones;
   (b) *Semiotic*, or social, with biological anchoring.

2. The arena of the immune reaction is contained within the skin; the arena for signal anxiety is normally between the perimeter of the Hediger 'bubble' and the skin of the organism, the former containing the latter.
3. Invasion of (a) is initially signalled by the immune response, of (b) by anxiety, with the latter serving as an early warning system for the former.
4. In evolution, (a) is very old, whereas (b) is relatively recent. There is a corresponding advance from a purely metonymic nexus to one perceived as causal efficacy.
5. Communicational errors occur in both processes, and may have devastating effects on the Self (1991c: 40).

Sebeok's separation of the apprehensions of the self – *immunological* and *semiotic* – is obviously crucial. The immunological self operates in a 'semiotic' fashion; the semiotic self operates in the most complex and potentially unanticipated ways predicated on a biological impetus.

Again, as with language, selfhood consists of a complex of semiosic-organic interactions. The recognition of the self is set in a semiotic context of cognition and sociality, yet it evolves in bodily processes. The continuity across realms of nature suggests, this time, that the phenomenon under scrutiny – different forms of subjectivity – grows out of other-reference: biologically. Petrilli and Ponzio (2005) have also developed this insight for a theory of culture and communication. Yet, this should not be grounds for reaching a determinist conclusion on the matter, whereby subjectivity is summed up as a mere by-product of nature, a uniform filtrate of basic cellular interaction. Biosemiotics has routinely avoided such determinism and has done so pointedly, through its stress on agency and its contravention of 'eliminativism'. This has not just taken place at the level of discussion of higher organisms; biosemiotics has also revolutionized understanding of the lower organisms by way of a semiotic framework. Moreover, this framework implies both cognition and sociality, taken broadly. For example, Hoffmeyer writes, "Even an amoeba is capable of choosing to move in one direction rather than another. It will, for example, generally gravitate toward the richest source of nourishment" (Hoffmeyer 1996: 48). In this way, what biosemiotics introduces into the study of nature is "intentionality", a difficult concept because, like agency, it comes from the world of culture whose very variegation seems to render its concepts inapplicable or not susceptible of translation (the other way, this time) into the realms of nature.

Within biosemiotics the debate about the definition of 'intentionality' continues (see Favareau et al., forthcoming); yet, Hoffmeyer is circumspect when he suggests that living entities developed into "intentional systems – subjects in a sense – because they had established channels for an integration of other-reference (through surface receptors) with self-reference" (Hoffmeyer 2010a: 31). The implication is that intentionality in humans is a biological characteristic, a continuity, rather than the extensions of a ghost in the machine. So, while humans and their culture are not simply a result of biological determinism, nor are they children of the environment as in so many debates about nature/nurture which have privileged the latter in what has been called "the Standard Social Science Model" (see Pinker 2003). Hoffmeyer and Kull write, "organisms do not passively succumb to the severity of environmental judgement, instead they perceive, interpret, and act in the environment in ways which creatively and unpredictably change the whole setting for selection and evolution" (Hoffmeyer and Kull 2003: 269–70). Agency, therefore, has to be seen in terms of the biological preparedness of surface receptors for other-reference or, at the level of culture, the 'answerability' to the other which is elided and occluded by the construction of an impossible self-contained identity (see Ponzio 2006a).

The dialectical nature of the biosemiotic perspective on subjectivity and agency is palpable and, furthermore, is carried over into one of its key arguments, encountered in the last chapter, regarding the nature of verbality and non-verbality, but inflected here in respect of the discussion of agency. Sebeok's theory of modeling (Sebeok 1988) is undoubtedly evolutionary in its bearing. Yet, as has been seen, the tri-partition of modelling also corresponds to the triadic thinking of Peirce (and Morris). Ontogenesis is seen to have its iconic, indexical and symbolic phases and components as well as phylogenesis, with the symbolic dimension harbouring a particular kind of importance for human thought and its foundation in a specific form of modeling. The three different kinds of modelling, it seems, broadly indicate different degrees of 'semiotic freedom'. What is important in respect of agency is that the third level, dominated by the 'symbolic', is where there is growth: "Symbols grow", wrote Peirce (2.302) and Sebeok (1977: 181) adds that Chomsky,

> whose way of envisioning the acquisition of language vividly recalls (as he is well aware) some of the ideas of Peirce on the logic of abduction – reconstituted, with the linguist's fruitful reformulation 'the creative aspect of language use' (Sebeok 1972: 6), this notion and turned it to challenging problems of modern linguistics.

Sebeok, observing the two traditions of sign study – semiotics and glottocentrism – notes that language is a relatively recent part of nature whose growth (in secondary modelling upwards) is "Lamarckian in style, that is, embodies a learning process that becomes part of the evolutionary legacy of the ensuing generations (1977: 182). Referencing Bateson, he then remarks that Darwinian and Lamarckian evolution co-exist in the human animal, making human evolution "not only a reconfirmation of the evolutionary processes which went on before man appeared on the scene, but continues as dual semiotic consecution that can scarcely be uncoupled in practice: one track language-free (zoosemiotic), the other language-sensitive (or anthroposemiotic)" (1977: 183). The importance of this statement for biosemiotics' cultural implications cannot be overestimated and it is notable that Sebeok does not see humans in evolution as *dependent* on language or 'constituted' as subjects by language but as *language-sensitive.* Humans hark back to their non-human ancestors but not in all respects. What separates human and non-human must be a special process of evolution which differs from the traditional accounts of Darwinism and Lamarckianism.

The discussion of these points runs the risk of taking a dangerous turn. While Hockett noted the importance of 'tradition' in passing on *a* language (as noted at the end of the last chapter), the genetic aspect in passing on language (without an indefinite article) has offered succour, in a roundabout way, to cultural exceptionalists.

The Lamarckian bearing of 'language', its apparent capacity to pass on acquired characteristics, is already cherished and valorised by cultural analysts even if they do not use this frame of reference. Indeed, this has been taken to be the whole of language, bracketing its modelling basis, and goes some way to explaining why Chomsky's work has had negligible impact (or at least take-up) in cultural and communication studies, despite the contemporaneous popularity of his wholly unrelated work on media and propaganda. The more nuanced account of language's phylogenesis and ontogenesis offered by Terrence Deacon and which has been discussed in Chaps. 1 and 3, above, reveals more about its role in human agency. "Deacon's scenario", writes Hoffmeyer (2008a: 294),

> points to the obvious possibility that language gradually changed in such a way that its grammar came to correspond optimally to what children immediately intuit. Since changing human brains is an extremely slow process, what happened instead was that language adjusted itself to the patterns of children's brains – and that the human brain the next time round adapted to the new linguistic challenges.

This is not evolution in the blind, eliminativist fashion; rather, it is agentive. Where language, evolution and agency is concerned, Deacon's thesis – with considerable detail from anthropology and cognitive science – is that symbols, characterized by growth, themselves characterize the quality of human linguistic communication. Symbols represent a leap beyond the indices and icons that animals use. Deacon does not use the word, nevertheless it is an 'abductive' leap in Peirce's terms, dependent not on relations to things but 'guessing' from a system in which the symbols are as likely to refer to each other as to things in the world. As Deacon points out, chimps can be taught to use a limited amount of symbols and, furthermore, as Hoffmeyer (2008a: 285) insists in respect of RNA and DNA, "the invention of *symbolic reference* was a theme in evolution a long time before humans appeared".

Yet Deacon argues persuasively that the capacity of humans to think symbolically was likely to have been nurtured by experience and learning of ritual. In its inversion of both Chomsky and Darwin, this perspective on evolution is Baldwinist. Nevertheless, Baldwinian evolution should not be understood, as it might be tempting for cultural analysts to do, as an evacuation of scientific principles in favour of free will. In their contribution to a volume on the 'Baldwin effect', Hoffmeyer and Kull are keen to avoid underestimating, in the fashion of some biologists, the intricacies of the project of nurturing. In comparing 'interpretation' of concentrations of adrenaline by a liver cell and 'interpretation' by a human observer of smoke, they sound a note of caution. They argue (2003: 261) that the

> liver cell is underdetermined to the extent that the contextual situation in which it finds itself does – if ever so weakly – enter the myriad relational connections making up for cellular control. For instance a diversity of cellular recognition processes are mediated by the same G-proteins, and different G-proteins can occasionally be utilized by the same receptor.

The cell has a kind of 'choice', exemplifying agency, through receptor and effector, in the liver. Yet, Hoffmeyer and Kull's stress on choice rather than determination, like Deacon's pointing to ritual and learning, should not be allowed to obscure a complementary point. Moreover, it is a point which corresponds with the workings of anxiety and the immune system. That is, that the human brain's interpretation and choice is *consanguine* with that of the liver, the former related to the latter by dint of endosemiosis and, to echo Sebeok, above, as "a reconfirmation of the evolutionary processes which went on before man appeared on the scene".

The stress on endosemiosis here is intended to avoid 'over-egging the pudding' of biosemiotics' findings regarding agency. To stress agency in a community where agency is dismissed or undervalued is one thing; to over-stress agency in a community that already over-values agentive action to the extent that it neglects or denies its existence outside of culture is another. It is important to remember, then, that some of the key principles of biosemiotics provide a bridge between the humanities and science, most manifest, perhaps in the movement of biosemiotics towards the humanities through its take on agency in the natural world. At the same time, this should not license repression of the fact that bridges are not necessarily built for one-way traffic. That is, with culture-friendly premises also come the demands of science. An example of this is Hoffmeyer and Emmeche's pivotal concept of code duality. Put briefly, they posit a code for action and a code for memory (Hoffmeyer and Emmeche 2007: 27) in life. Hoffmeyer adds that "Every life form exists as itself, i.e., as an organism of 'flesh and blood', and as a coded description of itself, the latter being lodged within the remarkable DNA molecules of which the genetic material is composed" (1996: 15). Organisms do not survive forever; instead, they pass on signs as 'versions' of themselves, making heredity into "semiotic survival" (1996: 24). Action in a lifespan is dominated by analogue signs, changeable and interpretable to different degrees; whereas "genetic memory works as read-only" (Kull 2007: 8). It is no accident, Hoffmeyer and Emmeche note, that the code for memory is digital; yet, *passing on* digital memory, can only be seen at the level of the population. It is not to be observed at the level of the individual (2007: 35). The broader point that code duality seems to underline is that digital codes have been imputed with an autonomous character when, in fact, their sphere of efficacy is limited. This would be generally music to the ears of cultural analysts who, as a group, have had little time for biological reductionism as a potential encroachment on cultural policy – in all its ramifications, from learning theories and practices to policing.

Still, code duality should not be taken as a rallying call to celebrate autonomy or, worse, individualism, either in the realm of genes or in the realm of analogue coding. Assessing the Dawkinsian perspective, Hoffmeyer and Emmeche oppose it to common sense, quotidian individualism which, in an interesting twist, is shown to overestimate the fixity and authenticity of putative innate agency over human fraternity and heritage:

> In our opinion this utterly reductionist view [the selfish gene] grants to the digital code far too much independence. The opposite may be said of the common sense conception according to which the individual is unique and should not therefore be considered a tool for anything but himself or herself. Everyone is the architect of his own future, as it goes. From a biological point of view this individualism would seem rather ungrounded. Individuals after all are mortal and without sexual reproduction they would not exist. Thus,

> individualism – biologically seen – tends to overestimate the independence of the analog phase of active life and underestimate the significance of the species history as reflected in the digital code of the common gene-pool. Both 'DNAism' and 'individualism' tend to blind us to the importance of the code-duality, i.e. the subtleties of the *translation processes* between analog and digital code (2007: 51)

Bearing upon the matter of translation once more, Hoffmeyer and Emmeche's observations suggest that while biosemiotics, because it is embedded within semiotics as a whole, employs the vocabulary of culture, it also entails the compulsion for the humanities to mount the bridge between disciplines to survey the persuasive picture of continuity across nature.

A further case in point is the idea of 'semiotic freedom' (Hoffmeyer 1996, 1998, 1999, 2008a; Hoffmeyer and Emmeche 2007) which Hoffmeyer calls "the inner core of organic evolution" (2010a: 31). The idea, certainly at first sight and along with the general positing of agency in nature, seems to move science that much closer to the qualitative concerns of culture. Hoffmeyer refers to the first living systems' development of an ability to anticipate or 'interpret' regularities in their surroundings as signs. It is a decisive moment in which, as Kull would argue, recognition, memory, categorization, learning and communication are aligned. Ultimately, the reading of such signs of regularity might have a bearing on the future of those living systems. As seen in Chap. 1, above, Hoffmeyer gives the example of "when a bacterium 'chooses' to swim upstream in a gradient of nourishment rather than tumbling around waiting for the nutrients to reach it" (2010a: 34), noting a "talent for anticipation" which would have started a tendency to systems with more and more semiotic freedom of this kind.

Another way of phrasing this configuration is "*interpretance*", the capacity of a cell, organism, species to distinguish parameters in its surroundings or its own interior and use them in regard to significance. Hoffmeyer does point out that the semiotic freedom of agents is very low at primitive levels and that it is a species property and not an organismic property (2010a: 35); yet, before anthropocentrists and glottocentrists jump to the conclusion that the human's semiotic freedom is of a totally different breed, he adds in a footnote

> Even at this level one cannot rule out individual semiotic freedom right away, though. A bacterium is a hugely complex and well tuned system of proteins and other components and although learning processes probably do not directly play a role at this level the bacterium is capable of changing its behavior by the active uptake of foreign DNA from bacteriophages (2010a: 35 n. 9).

For the most avidly anthropocentric, the attention to the complexity of semiotic resources in an organism and their extensions that is at the centre of semiotic freedom as a concept contrasts with the bleakness of natural selection (which culture studies, of course, have largely chosen to ignore in any case). It proceeds from a Baldwinist premise: "The trend toward increased semiotic freedom, and the consequent pressure for still better exosemiotic function, created pressure for still more endosemiotic refinement" (Hoffmeyer 1992: 111). The point of semiotic freedom – continuity of interpretance from lower to higher organisms – is its potential contribution to understanding such cultural preoccupations as adaptability, interpersonal relations, subjectivity, art, the good life, value and ethics, unsettling all those who would bluffly draw a line between humans and the rest of nature.

Semiotic freedom is also at the centre of *Umwelten.* However, before noting this point at greater length, it is worth cautioning, once more, against the temptation to equate 'freedom' with autonomy, teleology and progress. Indeed, the broad implications that cultural analysis needs to embrace from biosemiotics actually instance the ways in which there is much semiosis that is repressed in human ontogeny and elsewhere. In the Deaconian evolutionary paradigm for language, indexicality supersedes iconicity and, in turn, symbolicity supersedes them both. In addition, though, as Hoffmeyer (2008a: 290) attests,

> A mental inversion must take place that pushes the objects and events of the surroundings to the mental background in order to permit the establishment of a new systemic web of word-to-word (or more accurately, symbol-to-symbol relationships capable of imprinting meaning – one based on the network of relations between words and other words (symbols and other symbols) rather than on the more fixed and dyadic relations between words and reality – upon the flat indexical backdrop.

What Hoffmeyer is referencing in this quote is a kind of 'repression', although, as will be seen in Chap. 7, below, it needs to be discussed in terms of 'constraints'. Certainly, the phenomenon in question appears as an evolutionary supersession at the level of the species but something is lost at the same time nonetheless.

One of the key processes identified by biosemiotics through its nesting, embedding and sometimes constrained situatedness is 'semiotic scaffolding'. The construction work term 'scaffolding' had been adopted and developed in the work of the psychologist Jerome Bruner (1957, 1960, 1966) and interpreters of Lev Vygotsky such as David Wood (Wood et al. 1976) in relation to young children's building on already mastered skills in the process of learning. In biosemiotics, Jesper Hoffmeyer further developed the concept, generalizing it to cover the network of semiotic interactions connecting an organism with its *Umwelt*, facilitating its processes of perception and action: "The network of semiotic interactions by which individual cells, organisms, populations, or ecological units are controlling their activities can thus be seen as scaffolding devices assuring that an organism's activities become tuned to that organism's needs" (Hoffmeyer 2007: 154). This biosemiotic use of 'scaffolding' has several aspects. One is genetic assimilation – the idea that structures appearing in the lifetime of organisms may, over generations, become genetically coded, provided those structures give the organisms selective advantage. Here, the scaffold metaphor is stretched a bit – or used creatively, as it were: the scaffold is not taken down when the building behind it is finished, rather, the scaffold *becomes*, over time, part of the building itself. Another aspect of 'scaffolding' has to do with the articulation, subdivision, detailing of a process so that those process parts or aspects may receive a higher degree of detail control; the more sub-processes are rendered partially autonomous and hence controllable, the more probable is the safe and successful completion of the overall sum process. Simultaneously, the autonomy of parts may facilitate a higher degree of flexibility by means of making different combinations of parts possible. Still another aspect of much, if not all, 'scaffolding', highlighted by the metaphor, is its external, material aspect in relation to the single organism: many organisms do not simply exist in an otherwise unchanging, neutral environment; rather, their activity to some degree shapes and changes their Umwelt so that its affordances more easily allow for the organism to enact its activities. Finally, according to Hoffmeyer's argument, such scaffolding invariably has semiotic aspects: the piecing together of the semi-autonomous parts of a scaffolding has the character of meaning-bearing couplings as they support still more complicated versions of the basically significant perception-action cycle.

The scaffolding concept thus plays a major role in a biosemiotic worldview. Much research into cognition (for example, Donald 1991; Tomasello et al. 2005 and many others) has pointed to the fact that the strange and growing abilities of humans have emerged through intersubjectivity and the *co-evolution* of culture, language, and brain in relatively recent humanoid history. The brain is not to be conceived as a computing mechanism dictating motor actions and cultural interactions. Nor are culture and civilization any longer to be taken as mere icing on a biological cake already baked. Rather, culture and civilization have, at least since early development of language in hominids, if not earlier, fed back onto evolution. Thus, those humans who have been more able to learn, teach, and develop further language and culture have been favoured in the process of survival. This is also the view of the 'Baldwinian evolution' that biosemiotics has been instrumental in revivifying (Weber and Depew 2003). In this scenario, features such as the large human neocortex, the brain's linguistic circuits, hands able to grasp objects, and so forth, seem very likely to have co-evolved with human culture, communication and tool use. The interaction of these sets exemplifies, as it were, scaffoldings which have, over the course of generations, become part of the construction itself.

Peirce, as the father of pragmatism *and* semiotics, emphasized the externalization of signs, closely related to possible pragmatic action. Hence, for Peirce, externalized signs are not mere supportive devices; instead, they take on tasks which simply could not be performed by the brain alone:

> Again, the psychologists undertake to locate various mental powers in the brain; and above all consider it as quite certain that the faculty of language resides in a certain lobe; but I believe it comes decidedly nearer the truth (though not really true) that language resides in the tongue. In my opinion it is much more true that the thoughts of a living writer are in any printed copy of his book than that they are in his brain (7.364).

To put the matter another way, the author's brain is indispensable for writing the book – but the contents of the book as a whole were never once present in the author's mind; rather, the long and cumbersome process of writing constructs an artifice which contains thoughts and reasonings whose sum transgresses, by far, the online capacities of the author's here-and-now consciousness. This immediately is an offload function: the book remembers far more, and far more accurately, than the brain involved in its construction. But that is not all: having externalized an argument structure in a book chapter, the writer is free to take the results as new starting points, as scaffolds, for the next chapter – effectively constructing the book as a long, coherent argumentative arc which was never present to the author's mind in its entirety. Signs, in this way, are indispensable scaffoldings for humans in thought and action. This comes to the fore in Peirce's doctrine of diagrammatical reasoning – the manipulation and experiment with diagrams, externalized, in the imagination, or the two in combination – is taken to be central to thought and cognition (cf. Stjernfelt 2007).

The process, of course, is not limited to books (even if the technology of writing seems particularly important for the inheritance and accumulation of culture over generations). Indeed, the Toronto School after Innis and McLuhan has been dedicated to pursuing the scaffolding processes or extensions of humans in their mental and physical habitation of technology. Institutions, arts, crafts, infrastructure and technology form externalized scaffolds, moulding human behaviour in certain directions, affecting the bequests and reinterpretation of these scaffolds as well as the ongoing cultural selection between them, making possible their further development over generations.

The concept of scaffolding points strongly to co-evolution, giving the lie to a number of binaries. These include organism/environment and, for humans, language/brain and culture/biology. Indeed, the current chapter has witnessed how biosemiotics repeatedly shows the untenable constitution of certain binaries upon which some understandings of culture lie. The verbal/nonverbal is collapsed in primary modelling. The non-human/human binary is collapsed in scaffolding (where the human is not separated from its cultural artefacts) and in the general synechism (where the human is not separated from other collections of signs). Similarly, the binary of individual/collectivity, is also thoroughly undermined by synechism. In a related fashion, agent/subject loses some of its purchase as agency is shown to exist across nature and a kind of subjectivity or selfhood derives from reactions which take place at the level of the cell. The Cartesian opposition of body/mind is put to rest by endosemiosis. Underpinning all observations in this chapter is biosemiotics' abolition of the oblique slash between living nature/culture. The biggest binary of the lot, matter/mind is, of course, an ongoing project, although Deacon's (2012a) contribution represents a major landmark whose formulations are adumbrated whenever reference is made to the nesting of indexicality in symbolic reference and the exercise of 'constraint'. The only two-part entity that biosemiotics seems to leave intact is 'code-duality' and that is not strictly a binary opposition so much as a coupling through translation.

Like the various nature/nurture debates that have taken place throughout the years, the dialectic of inside/outside has troubled theorists of subjectivity. The two problems associated with such debates have been the means to achieve a balance and the extent to which the actions of an agent – variously formulated at different times as 'free will', 'individualism' or 'destiny', 'fate', and so forth – might play a major role. Throughout biosemiotics such dichotomies are given short shrift; 'nature/culture' amounts to a pointless detour, given that, for Sebeok and those that have followed in his path, the second part of the dyad is, once more, "that minuscule segment of nature some anthropologists grandly compartmentalize as culture" (1986a: 60). His comments on the semiotic self, once again demonstrate that selfhood and semiosis have an entwined heritage that ensures that semiosic products called 'culture' are just another part of nature. Certainly, the reiteration of this continuity should not just be passed over in this context. Bourgeois humanism, as Althusser shows (1969: 247), makes 'man' the principle of all theory and the human's products in culture must therefore be exceptional. Nixing that exceptionalism, as biosemiotics does, amounts to what can truly be called anti-humanism, in the same way as biosemiotics is anti-eliminativist. The theoretical anti-humanism of Althusser and his successors is dedicated to re-orientating humanist currents in philosophy and politics, such that 'the human' is no longer to be accepted as a sapient unified conscious with its own nature rendering it morally and intellectually independent of all determinations, no longer to be extolled for that independence which has unproblematically installed it at the centre of all considerations regarding the universe. Sebeok's biosemiotics is even more truly anti-humanist: it does not suggest that the human is determined by the capitalist mode of production; while that may very well be a fact, the more important point to recognize is that the human is, like all organisms, a repository of certain agentive functions within a set of determinations generally called 'nature' (see Cannizzaro and Cobley 2015). The anti-humanism in biosemiotics envisages humans within semiosis and within *Umwelten*.

In contemporary semiotic terms, humans do not pre-exist semiosis and then struggle when they are somehow "inserted" into it. Nor are humans the conscious creators of semioses by which they can exercise control and power. In an *Umwelt*, humans inhabit from the start the very signs that their sensorium allows them to promulgate. Humans cannot "get outside" semiosis and control it; along with other living creatures, they *are* semiosis. This corresponds with another anti-humanist plank of Althusser's work: his positing of ideology as a lived relation constituting concrete reality. For Althusser (1969: 233), the imaginary and the lived are in a complex interplay where ideology

> is a matter of the lived relation between men [sic] and their world. This relation, that only appears as "conscious" on condition that it is unconscious, in the same way only seems to be simple on condition that it is complex, that it is not a simple relation but a relation between relations, a second degree relation. In ideology men do indeed express, not the relation between them and their conditions of existence, but the way they live the relation between them and their conditions of existence: this presupposes both a real relation and an "imaginary", "lived" relation. Ideology, then, is the expression of the relation between men and their "world", that is, the (overdetermined) unity of the real relation and the imaginary relation between them and their real conditions of existence. In ideology the real relation is inevitably invested in the imaginary relation, a relation that expresses a will (conservative, conformist, reformist or revolutionary), a hope or a nostalgia, rather than describing a reality.

In this formulation there is an indication of the consonance of Althusser's conception of ideology with the insistence of biosemiotics that the sign is always a relation, but a relation oscillating between mind-dependent reality and mind-independent reality. Clearly, Althusser's insight into ideology as both "lived" and a "relation" was groundbreaking, more so than the representational paradigm which grew out of the view of ideology as involving subjects locked in "imaginary" relations. Yet biosemiotics has a more consistent approach as a result of focusing on ideology's instruments and its effects in terms of that which more broadly constitutes them: human semiosis. Thus, biosemiotics discerns that the existence of the human *Umwelt* itself consists of constant fluctuation between what is mind-dependent and what is mind-independent well before power relations effect an 'imaginary' field. This is not to downplay power relations; rather, it is meant to show that they have a broader foundation in 'objective reality'. Deely (2009a: 243–4) shows how objects are 'objects of experience' by definition; they require a subject and the 'objective world' is a world dependent on experience (rather than the existing world 'outside' experience that common parlance assumes):

> The most important point in the social construction of reality, no doubt, occurs in the political order, when the semiotic animals sit down together to try to decide how to govern themselves, how to decide what is to be permitted and what not permitted in social behavior and arrangements. Thus the constitution of a state, for example, the document, I mean, which details what the arrangement shall be for a given human community, is a prime example of a purely objective reality which can yet be realized in the subjective order of living and interacting individuals. Reality as we experience it is neither purely objective nor purely subjective nor purely intersubjective, but rather a constantly shifting mixture and proportion of all three – a mixture and proportion of which it is not at all easy (perhaps not even fully possible) to keep track.

Furthermore, in biosemiotics the reference to the 'real' – to the 'thing' beyond the 'object' – is an index which is nested in human symbols, rather than something imaginary harbouring something real as it is for Althusser. Plus, of course, human semiosis is embedded in semiosis in general, including that of animals and plants, meaning that biosemiotics posits a natural subject in a much more extensive set of semiotic determinations than those thrown up by civilization.

This also means that what is 'other' in biosemiotics is conceived in a more far-reaching fashion than in other theories of subjectivity. Customarily, the other has been someone – or, occasionally, some thing – else. It is often another gender, another sexuality, another ethnicity, another nationality, another culture in general. In posthumanism, the other has often been a non-human animal, a machine or an organic entity augmented by prostheses such as a cyborg. Invariably, the other is considered in such formulations in order to demonstrate the 'undecideability' that arises in attempts to pin down the nature of the human and subjectivity. As has been seen, biosemiotics is not prone to such squeamish prevarication; it is committed to revealing what it is to be human and what it is to be a natural subject. Furthermore, it is in an enhanced position to do so – not by leaving 'undecideability' as the final point on the journey but in taking it as the starting point ententionality. As Deacon (2012a: 534–5) declares,

> I believe that human subjectivity has turned out not to be the ultimate 'hard problem' of science. Or rather, it turns out to have been hard for unexpected reasons. It was not hard because we lacked sufficiently complex research instruments, nor because the details of the process were so many and so intricately entangled with one another that our analytic tools could not cope, nor because our brains were inadequate to the task for evolutionary reasons, nor even because the problem is inaccessible using the scientific method. It was hard because it was counterintuitive, and because we have stubbornly insisted on looking for it where it could not be, in the stuff of the world.

Deacon sees subjectivity as emerging from the teleodynamic processes of the brain. There are neuronal and energy substrates; but the nature of their constraints mean that there are processes that they do not actualize. Thus, “We are what we are not: continually, intrinsically, necessarily incomplete in our very nature” (Deacon 2012a: 535). In the face of such incompleteness, whatever is other to, outside a process or “not there”, assumes particular prominence in generating a sense of self.

In its ententionality, the other that selfhood encounters is as natural as everything that is encountered in an *Umwelt*. As Deely (1994: 15) notes,

> Otherness, then, in the sense of a physically opposed other, an alternate subjectivity, not thematically seized upon as such, to be sure, but given as such nonetheless in the actual encounter, appears as an *element* of experience: that element which is irreducible to experience of it. Otherness, more precisely, *at this level*, is that element in the experience as a whole which demonstrates that experience as a whole is not reducible to the existence of things, and the existence of things is not reducible to our experience of them.

The experience of the subject, although potentially very rich, is not an experience of things but largely a living through objects. That those objects have been invariably organized in such a way as to maintain particular kinds of power relations has no doubt been true for a long period of the history of hominid existence. Yet it is not the only result of objective experience, nor is it inevitable; for what biosemiotics finds to be ‘other’ for the self/subject in the natural world is something much more significant than another member of a species with, say, highly specialized sexual practices. As Sebeok’s investigations as well as the theory of semiotic scaffolding suggest, the other is everything and oneself.

Chapter 5
Ethics Cannot Be Voluntary

In biosemiotics, ethics is to be considered as arising mainly from three features of the human *Umwelt*. The first is in the displacement capacity of language: the possibility of signifying, in remote fashion, other times and places, things that have not yet happened (fictions), as well as anticipation and ideal scenarios based on an evaluation of current ones. This latter displacement capacity is ethics. It is probably closely related to other faculties of anticipation as well as the displacement projections mentioned here. The second is in all the aspects of the *Umwelt* that contribute to experience, including experience of pleasure, pain, sadness, happiness, well-being and so forth. The third is in the specific experience of otherness that accrues in the human *Umwelt* that was discussed in the last chapter. Although these features are sometimes alluded to in passing in the biosemiotic literature (honourable exceptions include Deely 2016 and Weber 2016), analyses of the ontology of ethics in biosemiotics have been somewhat circumvented. Instead, articles on ethics in biosemiotics have tended to immediately jump to discussions of ethical and moral questions that might be approached in a biosemiotic frame, such as the value accorded to different inhabitants of the biosphere (Tønnessen 2003; Beever 2012; see also Beever and Tønnessen 2016).

The current chapter is concerned more with the implications of how biosemiotics formulates the possibility of any ethics ever existing. That implication, in outline, is that ethics does not arise from the sound moral judgment of a rational, unified consciousness. Rather, it is a product of the conditions of the human *Umwelt* that also bring forth the human as different in kind and degree from other animals as well as the human as a natural subject discussed in the preceding chapters. Ethics, in the context of biosemiotics, must have consanguinity with the rest of the biosphere. This is in marked contrast to a picture in which ethics emanates from an invented human project. There are good reasons for a biosemiotic formulation of ethics to take issue with such a scenario. The idea of 'ethics' as a moral system, an idea which has developed from the early seventeenth century onwards, contains a basic contradiction in that it implies both a programme for behaviour and the will or agency to produce, adhere to and reproduce that programme. The latter, 'willed', ethics has

P. Cobley, *Cultural Implications of Biosemiotics*, Biosemiotics 15,
DOI 10.1007/978-94-024-0858-4_5

roots in the Greek *ethos* and its concern with matters of character and the personal. In the late twentieth century poststructuralism rightly cast suspicion on the notion of ethics but also tried to re-draw its programmatic aspects by calling for an ethics characterized by 'openness' to the other. Even here, the idea of openness itself suggests a programme of initiative activity or will by which dialogue can be achieved. It remains allied to individualism, rationalism, Cartesianism and, among other things that are not 'isms', the potential for local and global failure.

Clearly allied to the conception of will, certainly recognizable in the contemporary Western social formation and, unsurprisingly, associated with that poststructuralist moment, is the sense of ethics as a phenomenon in discourse. The programme of ethics is repeatedly framed as discursive, often appearing in institutional space precisely as a written code, and the grounds upon which ethics can be challenged or adjusted are likewise discursive ones. Indeed, much of the problems associated with the (lack of) efficacy of ethics, particularly in the era of multiculturalism and tolerance of the other, are derived from the belief in the putative discursive nature of ethics. The idea that many of the determinants of human life are 'constructed in discourse' has been a powerful one during the last 30 years, especially in relation to understanding subjectivity. Calvin O. Schrag's (2003) positing of 'communicative praxis', for example, constitutes an important logical argument regarding the contiguity of *communication* and *action*, showing how such enterprises as ethics – willed and programmed – are necessarily conducted through discourse.

Other sources of the 'discursive imagination' have lent weight to the perspective in which human affairs and the effecting of change in human affairs are determined by the vicissitudes of discourse. These include the 'linguistic turn' in social thought, structuralism, poststructuralism and other glottocentric perspectives (see Chap. 2, above). For this perspective, it might be tempting to cite the opening words of the Gospel of John: "In the beginning was the Word …" In biosemiotics, it would be more apposite to suggest that "In the beginning was semiosis …" Nevertheless, both would have their own resonances in opposition to a Trotskyite position in left politics which bears very much on the issue of ethics. In the 1923 work, *Literature and Revolution* (1992), Trotsky castigated the "pure art" claims of Russian Formalism (and, by extension, 'formalism' in general) in contrast to the thoroughly utilitarian perspective on human semiosis evinced in materialist dialectics. He writes (1992: 41):

> The formalist school represents an abortive idealism applied to the question of art. The formalists show a fast ripening religiousness. They are followers of St. John. They believe that 'in the beginning was the Word' But we believe that in the beginning was the deed. The word followed, as its phonetic shadow.

Trotsky is at pains to point out that artistic communication has only ever been wielded through class interests. In the process, he predicts that while the art of the slave-owning class dominated for thousands of years, that of the bourgeoisie took over only for a matter of *hundreds* of years (1992: 44) and a proletarian art might be possible within decades. Yet, crucial to this would be a proletarian intellectual vanguard, a voluntarist movement which would complement the vanguardist success of

November 1917. The vanguard, presumably would carry out 'deeds' – as though 'deeds' had no semiotic component and lay outside the actions of signs.

Voluntarism and vanguardism have constituted a thorny problem for Marxism, pre-dating even Lenin (see, for example, Gouldner 1980). In fact, they have also informed political debate beyond Marxism. In the Leninist mould, the vanguard constituted an absolutely necessary corollary to a forthcoming revolution. It entailed revolutionary theory but also, and especially later with Trotsky, a voluntarist *push* for the implementation of that theory. On the one hand, vanguardism is envisaged as arising from the proletariat; yet it has been associated in the history of Marxist states with authoritarian attempts to force revolutionary conditions. Like its philosophical counterpart, voluntarism in politics emerges through acts of will and, as such, complements the vanguardist impulse. Extreme voluntarism is sometimes viewed as an impetus which might 'corrupt' the organic basis of real political struggle. If it needed saying, voluntarism involves agency but instantiates the kind of Cartesian rational free will that biosemiotics has eschewed as an explanatory principle.

Vanguardism and voluntarism are mentioned here because, with any political impetus, including one grounded in ethics, there is always bound to be a degree of 'will' invoked. In a post-Marxist environment, the extent of that voluntarism will be derived from the theoretical basis upon which it is predicated. This predicament, apparently, is the one that is faced by a penetrating biosemiotics-inspired theory of ethical principles: 'semioethics'. Semioethics derives from proposals put forward by the semioticians, Deely et al. (2005), who have, together and separately, effectively carried the torch of Thomas A. Sebeok's teachings with aspects of this project. They proceed from the observation that the human is the only 'semiotic animal' (see Chap. 3, above) whilst all organisms are bound up in semiosis; thus only the human possesses a self-consciousness about the signs s/he uses. The crux of semioethics is that the self-consciousness of the human constitutes an absolute compulsion to all others in the environment of semiosis, a duty of care not just to humans but to all living things. The reasons for this are revealed in earlier work on dialogue by Petrilli and Ponzio (1998). The liberal conception of dialogue, they find, involves an impulse in which there is a decision to recognize the difference of the other and to 'grant' it respect. It features an ethics in which the human makes the effort to mediate between the positions of oneself and another, paying attention to the relation of another to one's own position. It commonly comprises two entities coming together in compromise, consensus and functional agreement. Such a conception involves unproblematized human agency and is irredeemably anthropocentric. It is not compatible with the depiction of the ineluctable encounter with the other that Sebeok identified in so lowly a sphere as the cell (see Chap. 4, above). Petrilli and Ponzio insist on going beyond the liberal notion of meeting others halfway, negotiating and compromising. Indeed, their conception *opposes* such agentive programmes by recognizing in dialogue a compulsion and demand rather than self-identified good will. Such a framing of dialogue is to be found, too, of course, in Bakhtin; as Petrilli and Ponzio succinctly state:

For Bakhtin, dialogue is not the result of an initiative we decide to take, but rather it is imposed, something to which one is subjected. Dialogue is not the result of opening towards the other, but of the impossibility of closing (1998: 28).

Ponzio (2006a: 11) attests that a rigorous conception of dialogue should not be self-justifying and seen in the service of mere self-affirmation as many theories of identity assume, wittingly or unwittingly.

Meanwhile, Deely (2005a: 11, 26) demonstrates that the formula of the human as 'semiotic animal' is not new and was ripe for development. It first appeared in 1897 in the work of the German mathematician, Felix Hausdorff (writing as Paul Mongré; see Mongré 1897: 7). Deely suggests that the designation and the conception that goes with it supersedes the modern notion of the human as *res cogitans*. As such, the human is not so much the thinking being of Descartes, but the being who comes "to realize that *there are* signs distinct from and superordinate to every particular thing that serves to constitute an individual (including the material structure of a sign-vehicle) in its distinctness from its surroundings" (2005: 73). He adds (2005: 75),

With the definition of the human being among the animals as the only *semiotic animal*, that is to say, the only animal capable of recognizing that there are signs (as distinct from their practical recognition and use) and capable of developing accordingly a semiotic consciousness of the radical role played by signs as well in the inescapable realism of animals as in the growth of all experience and of human understanding in particular, with its symbols everywhere in culture, we locate ourselves along a way of signs which leads 'everywhere in nature, including those domains where humans have never set foot'.

In consonance with what has been discussed in the preceding chapters of the current volume, the three important points in this formulation, then, are

that the human needs to be considered in his/her kinship with other animals a sign-user in common with all life-forms, including those that humans may not have even encountered;

that the human, thus conceived, is not defined by the power of thought as in the 'modern' paradigm, but, instead, by its existence within the whole web of semiosis, including endosemiosis;

and, importantly, that the attribute that the human does *not* share with other forms of life on the planet – the attribute that makes the human distinct – is consciousness of the existence and use of signs.

For Deely, the idea of the semiotic animal especially marks the human ability "not merely to distinguish things within objectivity, but further to explore them as they are in themselves" (2005: 58). This is the key point in semiotics' recovery of the *ens reale* (see also Deely 2003). Yet, the human consciousness of signs also entails that humans are in a position

to make the adjustments necessary from the metasemiotic standpoint for the well-being of human life *precisely in its dependency* upon the semioses which link human animal within the signosphere with the forms of life and semiosis by which the biosphere as a whole and the physical environment form 'nem'. (Deely 2005: 58)

In the concept of "adjustments" in this quote, the roots of a semioethics, as well as, seemingly unavoidably, voluntarism – the action of human will on signs – is visible.

Despite this lapse into rational will, the concept of semioethics can be maintained at its inherently most instructive by reference to its lineage in Sebeok's global semiotics and, in particular, Sebeok's embrace of the "one system" to which Deely refers. In 'The evolution of semiosis', Sebeok (2001a: 29–30) cites Lovelock's 'Gaia thesis', noting that

> All living entities, from their smallest limits to their largest extent, including some ten million existing species, form parts of a single symbiotic ecological body dubbed *Gaia* ... Should a view, along these lines, of a modulated biosphere prevail, it would in effect mean that all message generators/sources and destinations/interpreters could be regarded as participants in one gigantic semiosic web ...

In this light, what might be considered 'care of the self', can only realistically proceed from a 'care of others', where 'others' must mean the entirety of the semiosphere. It is in this sense that Petrilli and Ponzio's semioethics delineates not just a limited 'responsibility' but an "unlimited responsibility" to "all of life throughout the entire planetary ecosystem, from which human life cannot be separated" (2005: 534). Furthermore, central to their semioethics is the theorizing of otherness. For them (2005: 39–40), Levinas, Bakhtin and, crucially, Peirce, reveal that

> Otherness is inherent in the sign and at the same time the precondition for the sign's capacity to transcend itself. Signs – or better, signifying routes generated by the relations among signs in the macroweb of semiosis, or semiosphere – emerge from the tension between determination and indeterminacy; between a particular configuration of the sign and its continual displacement, transformation and deferral to the other; this other being both imminent to the sign and external to it, transcendent with respect to any given instance of semiosis. The other – this surplus or excess – prevents the sign totality from closing in on itself and thereby invests it with the character of openness and potential for creative generation. Openness or detotalization of the sign totality is the precondition for questioning and criticism, for the possibility of evaluating the operations of the 'mind', of semiosis, as good or bad.

Otherness, therefore, is thoroughly grounded in the sign. This implies that human will is, at the least, mediated – an agency that is compelled into compromise with circumstances. Yet, Petrilli and Ponzio (2005: 549) insist that "the entire planet's destiny, in the final analysis, is implied in the choices and behaviour of human beings". Moreover, they risk introducing further voluntarist overtones which seem at odds with the grounding of responsibility at the level of the sign, suggesting that "semioethics can be considered as proposing a new form of humanism" (2005: 545). Although Petrilli and Ponzio point out that their semioethics comprises a specifically Levinasian "humanism of alterity" (2005: 546) in contrast to liberal humanism with its associations to bourgeois individualism, the cultural implications of their biosemiotic casting of ethics need to be drawn out with reference to the more anti-humanist aspect of their project. Put another way, the idea of *commitment to* the other is to be replaced by the *compulsion from* the other to better understand the reformulation of ethics that biosemiotics implies.

Both 'commitment' and 'compulsion' are forms of what might be called 'motivation'. 'Compulsion' includes the coercion enforced by a normative programme or governing force, but here it is couched with reference to the call of the other. It has been argued elsewhere (Cobley and Randviir 2009) that the variegated analysis of anthroposemiosis that makes up the domain of sociosemiotics is characterised by attempts to understand the 'motivation' in the relations that make up the sign. Drawing on a seminal article by Kress (1993), sociosemiotics is described as an endeavour which untangles the relations between sign users, their circumstances, history and the materials that they use to make signs. These relations require elucidation because of the ways in which they are shrouded in opacity and because they obtain, frequently, at points when 'motivation' could not be revealed in its pristine state to an observer. Now, if this is the case, then one can easily deduce that the humanist imperative in respect of signs, which misconstrues the nature and limitations of agency in relation to sign-making, re-casts motivation as an entirely voluntarist affair. This is precisely symptomatic in the liberal conception of dialogue – a conception whose poverty Ponzio and Petrilli's work, individually and collaboratively, has been instrumental in demonstrating – where contact and 'communication' with the other is the result of a choice, disposition or other individual act. For many, dialogue retains the vanguardist tinge whereby, in the common phrase, one reaches out to another or, in sociosemiotic terms used here to supplement biosemiotics, where the relations of motivation between signs and their users is supposedly subject to an act of will. Again, Schrag (2003), for example, has a typical liberal conception of dialogue as intersubjective consensus in discourse.

Thus far in this volume, biosemiotics has been identified as driven by 'anti-humanist' conceptions. Although, as seen in the last chapter, biosemiotics surpasses the anti-humanism traditionally associated with some areas of post-Second World War Marxism, the latter is still revealing in relations at issue in the current chapter. Althusser, for example, in delineating "Marx's scientific discovery" (1969: 227), notes that the residues of humanism always need to be treated with care:

> Strictly in respect to theory, therefore, one can and must speak of *Marx's theoretical anti-humanism*, and see in this *theoretical anti-humanism*, the absolute (negative) precondition of the (positive) knowledge of the human world itself, and of its practical transformation. It is impossible to *know* anything about men [sic] except on the absolute precondition that the philosophical (theoretical) myth of man is reduced to ashes. So, any thought that appeals to Marx for any kind of restoration of a theoretical anthropology or humanism is no more than ashes, *theoretically*. But, in practice, it could pile up a monument of *pre-Marxist* ideology that would weigh down on real history and threaten to lead it into blind alleys (1969: 229).

Althusser does suggest that 'humanism' has its uses, as an idea and in ideology (1969: 231); but he is absolutely forthright about the need for thoroughgoing anti-humanism in theoretical work of the kind that semioethics might be taken to be.

It can also be argued that the Althusserian call for anti-humanism is coterminous with Sebeok's eschewal of humanism in the work that informs current semioethics. Sebeok traced his intellectual, semiotic lineage back through Peirce to Locke and, ultimately, Hippocrates; he was in no way moved by the humanist appeals of Condillac and other Enlightenment thinkers. In fact, his concern with endosemiosis and the immensity of the web of semiosis in general disqualifies the absurdities of much humanism from his thought. In theoretical terms, communication among humans amounts to pretty small beer. Sebeok's (2001a: 14–15) call for a consideration of the "staggering" amount of semiosic transactions in the human body puts interpersonal communication into perspective. Moreover, the communications to be found in the human body are merely extensions of the kinds of communication carried out by the earliest, and most enduring, organisms on the planet – bacteria. There is no doubt that communication between humans, particularly in the formation "communication-production", the profit-making imperative of global communication as identified by Petrilli and Ponzio (2005), has assumed a crucial position and has become, potentially, disastrous for the planet. Yet, where theory is concerned, there is a need to adhere to the larger picture of semiosis that Ponzio and Petrilli attempt to present, and there is a need, then, to maintain semioethics as an anti-humanist perspective in biosemiotics.

Maintaining anti-humanism in theory is not easy, of course. Althusser's successors have attempted to do so with varying degrees of outcome. One of the most celebrated of recent attempts is the project of Alain Badiou which, as well as potentially informing biosemiotics also presents actions that are best understood within the wider perspective that biosemiotics facilitates. His *Ethics* (2001) constitutes a summing up of his position that, although embracing human action and agency, rejects humanism and voluntarism. This rejection revolves around the co-ordinates of new situations, what Badiou names 'the event' and the fidelity that such events demand and which some humans can meet when they are *seized* by them. Baldwin (2004: 1) gives probably the most concise summary possible:

> The murmur of something new coming to happen interrupts the norm of a situation – that is, an event occurs. Those who correctly investigate the consequence(s) of the event and maintain fidelity to this interruption are subject to, and produce, a truth resulting from that event. This truth transforms existing knowledge and is universal – intelligible for everyone.

Badiou locates 'events' in a number of anthroposemiotic spheres: in explicitly political developments; in the act of falling in love; in the sphere of cultural practice. Likewise, fidelity will involve, respectively: continued commitment to a political idea; commitment to a relationship (that all one's friends might say is getting out of hand); and consistent faith in an artistic vision and practice (Badiou gives the example of Haydn as a revolutionary innovator – see Cobley 2004).

In the idea of 'the event', Badiou's work amounts to a sustained attempt to reconfigure what might be thought of in other political circles as voluntarism. Implicitly, 'fidelity' – which is fidelity to an event – supplants the role of the will and inverts the relation of will exerting itself on a situation. This inversion – or its opposite – should be familiar to semioticians, particularly as an analogous inversion has been discussed by Sebeok on numerous occasions. In 'Looking in the destination for what should have been sought in the source' (1979c), a classic essay in scepticism, Sebeok listed some ways in which scientists should "be ever on the lookout against deception, but beware, above all, of self-deception" (1979c: 95). In reference to this, the case that runs like a thread through Sebeok's work is that of Clever Hans, the 'intelligent' early twentieth-century horse who 'deceived' observers. The lesson of Sebeok and Rosenthal's (1981) collection on Clever Hans is, perhaps, above all, that deception is not just unwitting, but often "witting", in the sense that the deceived collude in their deception. They are often *willing* to be deceived at the outset, whether they believe what they witness to be the case or not. Stage magicians rely for the success of the majority of their tricks on the willingness of participants to be deceived.

Sebeok's observations on witting/willing deception find an interesting complement in Badiou's attempts to invert the relationship of will to situation. To put it another way, the *source* is responsible for the outcome observed in the *destination*, although all eyes are usually on the latter. Likewise, the *event* creates the *fidelity*, the latter of which seems to interested eyes to be rather a voluntarist act. Will, therefore, has to be uncovered; it has to be revealed where deceptions occur. Even in Badiou's problematic ethics, it is clear that deceptions between event and fidelity take place. Badiou cites the case of Heidegger, an intellectual who, in Badiou's (2001: 73) estimation, mistook Nazism for a revolution and thus fell prey to a simulacrum. In the same way, those who have suffered a recent family loss are usually the most suggestible to 'mediums' or 'spiritualists' in the way that Sir Arthur Conan Doyle was (see Stashower 2000). Such individuals are far too *willing* to participate; their will overcomes their ability to see the source, let alone look at it searchingly.

It should come as no surprise, then, that there is *willing* of this kind in political projects, even if it is not always couched in such obviously sceptical terms. Any political project which foregrounds its objectives as immediately achievable by will, is prone to utopianism, voluntarism and, sadly, disappointment. It is for this reason that pursuing ethics through willed political projects is so difficult. Engels (1946 [1886]) recognized this and noted, in what was a socio-political discourse but could quite easily have been a biosemiotic one,

> That which is willed happens but rarely; in the majority of instances the numerous desired ends cross and conflict with one another, or these ends themselves are from the outset incapable of realisation, or the means of attaining them are insufficient thus the conflicts of innumerable individual wills and individual actions in the domain of history produce a state of affairs entirely analogous to that prevailing in the realm of unconscious nature.

For Engels, expectedly, it is the vagaries of history which dictate the ground for political action or ethics. The biosemiotic concept of ethics is grounded in antihumanist principles which echo Engels, but they do so at the level of life, or living nature

In semioethics, the distinction between the realms of law and life would seem to be of considerable importance. Already, Petrilli and Ponzio have adumbrated the distinction in their discussion of how "human life in all its aspects has been incorporated into the communication-production network" (2005: 478). In this, they offer an analysis that is similar to that of Agamben (2005) who suggests that the modern political situation arises from a profound contradiction. Since at least the state of siege characterising the Paris commune, where traditional laws were suspended, the modern state has been built on an uncertain terrain between living being and law. This terrain has largely been hidden from view, preventing a proper appraisal of the differences between public law/political fact and life. For Agamben, the modern state has come to exercise its rule as though there is a perpetual 'state of exception' that necessitates the shelving of constitutional principals and, indeed, now constitutes the very being of the state. Petrilli and Ponzio (2005: 478) show that global communication, with its constant shifting of the goalposts, has likewise transformed the human experience with serious implications for

> development, well-being, and consumerism, or underdevelopment, poverty, and impossible survival; for health or disease; for normality or deviance; integration or marginalization; for employment or unemployment; for transfer of people functional to the workforce (which is characteristic of emigration) or transfer characteristic of migration, where the request for hospitality is denied; and for exchange and trade of legal merchandise or traffic in illegal merchandise, be it drugs or non-conventional weapons or human organs.

Agamben offers a further complement to Petrilli and Ponzio's explication of the domains of life, noting that the ancient Greeks did not rely on a single term to express 'life'. As Agamben points out, for them, two terms were in play: *bios*, "which indicated the form or way of living proper to an individual or group" and *zoē*, "which expressed the simple fact of living common to all living beings (animal, men, or gods" (1998: 1). Agamben further notes that simple life (*zoē*) is excluded from the Polis in the classical world.

The distinction and its consequences would seem to be of paramount importance for semioethics. In the analysis of Petrilli and Ponzio, there is a clear sense of the parlous effects on well-being of capitalism's omniscient manipulation of *bios* through communication-production. Additionally, there is a definite concern with bare life or *zoē*: human organs are undoubtedly a bare life constituent caught up in the struggle for biopower. Furthermore, Petrilli and Ponzio refer to *zoē* indirectly in their comments on the need for care of the planet and ecosystem in general: it is clear that the state repeatedly attempts to devour *zoē* and *bios* as one. Yet, there is good reason to place the *bios/zoē* distinction at the forefront when analysing the state's constant attempts to subjugate life as a general category. The state's superintending of *bios* is inevitable: it is associated with the *political* existence of humans in groups, where 'life' is automatically elided to the 'good life' sought by the Greeks (Agamben 1998: 7). As such, *bios* is in the seemingly ineluctable position of maintaining life insofar as it appears to best serve human groups. However, humans may be best served by a *bios* that is thoroughly predicated on *zoē*. This may be a more cogent formulation of Petrilli and Ponzio's (2005: 478) reference to the fate of "human life in all its aspects" since it clarifies the distinction in the quote from Deely, above, in which "the well-being of human life" is precisely dependent upon the animal- and plant-inhabited semiosphere. Whereas the state conflates *bios* and *zoē* in order to devour life, semioethics might insist on their distinction as a crucial means of drawing attention to the cul-de-sac of anthropocentrism. In the sphere of ecological debate this distinction has already been made in the opposition of anthropocentric and biocentric views on the environment, although it has occurred, significantly, without the benefit of the global understanding afforded by biosemiotics (Taylor 1986; Stenmark 2002). Anthropocentrism, unsurprisingly, is intimately related to 'will' in its placement of humans alone as prime movers on earth;

semioethics places humans among their cohabitants of the semiosphere (coterminous, of course, with the biosphere) and must therefore remain suspicious of untrammelled will.

Yet, the distinction between *bios* and *zoē* is not one that is utilised in biosemiotics. One reason for this has been discussed in the previous chapter and in relation to other issues in preceding chapters. It is that the human *Umwelt*, as a concept, already implicates a continuity of life in respect of the human as species and subject. The state may conflate *bios* and *zoē* for political purposes, to maintain particular kinds of power relations. However, beyond the recognition of a relationship of dependency between the non-human and human aspects of the biosphere, the distinction has little explanatory benefit for biosemiotics in general and for its discussion of ethics specifically. What is most important is that the life of the group involves self/other relations that exist across the habitation of any *Umwelt*. In the human *Umwelt*, with its potential for infinite variegation, the human subject is always prone to encounter the other in everything and in his/herself, not just in particular groupings. In recent liberal evaluations of the ethical relation of the subject to the group or community, Charles Taylor, Alasdair MacIntyre and others conclude that group life sets preconditions. The individual subject therefore is likely to realize her/himself adequately insofar as s/he meets those preconditions through shared values. Interestingly, Peirce, so often congenial to biosemiotics, famously prevaricates over the worth of ethics, sometimes seeing little to recommend it as a normative science (e.g. 1.573 ff) but later considering it the prerequisite to any science of logic (e.g. 2.198). Despite a fair amount of ink spilled on the matter by commentators, what does seem clear is that Peirce was extremely suspicious of conceiving ethics as a scientific positing of the moral values of a community and this tallies with his extreme disdain for the idea of the unified individual. Biosemiotics shares this suspicion and disdain. Ethics in biosemiotics, because it arises from the displacement facility of humans, might be malleable and manifested in a number of ways; ultimately, however, it is a nested product of the continuity of semiosis.

Given the foregoing misgivings about voluntarism, it seems that ethics, even with the insights of biosemiotics, might be severely restricted in its influence. It may be part of a humanist strategy in which individuals fulminate about the injustices and abuse inherent in late capitalism while simultaneously being prevented by its theoretical parameters from being involved in action. It may be a heuristic in cogent analysis of iniquities while only being able to say what should be done. There seems to be little sense in placing biosemiotics in the role of a social or ecological advocate in order to say simply that action is taking place. For example, an obvious response in such circumstances – to injustice and inequality on the one hand and ecological imbalance and human impingement on biodiversity on the other – would be policies of tolerance towards other peoples, a pursuance of multiculturalism and a local voluntarist impetus in respect of the environment. Yet, ideological imperatives of this kind have been such an integral feature of social democratic governments in the West during the late 1990s and early 2000s – the same governments, of course, who waged wars, exploited Third World labour and favoured fuel production over long-term environmental legislation – that it would be wise to be suspicious of them (see, *inter alia*, Barry 2000; Kelly 2002; Alibhai-Brown 2000; Bloom 2003).

This, along with Peirce's suspicion about moral programmes, should indicate why a biosemiotic conception of ethics adds something different from well-meaning policies that turn out to be harmful. It is worth noting that Sebeok's teaching emanates from a formidable figure who was far from being a fellow traveller of left causes; yet so seized was he (in Badiou's sense) by scientific and global principles that his fidelity to them provided semiotics with a template for thinking through and instituting a plethora of research projects that promise to bring so much closer the possibility of actions based on a biosemiotic ethics. Ponzio and Petrilli, while conversely more sympathetic to left causes, have also maintained fidelity in the global vision of semioethics: in the face of demands made by dogma that is sometimes to be found on the left, they have gone at least some way to bringing voluntarism and vanguardism into question. For many, semiotics and biosemiotics can be pejoratively taken to be 'apolitical' because of this fidelity. To assert as such, of course, is tantamount to saying that semiotics is politically conservative. For those who are of this view, it appears that semiotics has no socio-political bearing and the only way in which it can be co-opted is through those desultory attempts at what Barthes (1977b), in dismissing some of his earlier work, called 'mythoclasm' (see Chap. 3, above). More accurately, perhaps, what these critics of semiotics deplore is that semiotics is not voluntaristic.

There is an echo, here, of the 11th thesis on Feuerbach where Marx remarks that philosophers have only interpreted the world in various ways and that the point is to *change* it. Marx, of course, does not propose that this can be achieved through straightforward individual voluntarism in an ethical mode, such as being nice to people or recycling individual waste – humans, *as a collective*, will have to decide to do such things by themselves. Nor did Marx (or Engels) ever envisage a communicative praxis in which people speak 'correctly' in a political sense. Likewise, Petrilli (2005: 43–44) points out that

> Semioethics does not have a program to propose with special aims and practices, a decalogue, a formula to follow more or less sincerely, more or less hypocritically. From this point of view, semioethics is alien to *stereotypes* as much as to *norms* and *ideology* (Ponzio 1992, 1993, 1998). Semioethics proposes a *critique* of stereotypes, norms and ideology, of the different types of values as described, for example, by Charles Morris in his various writings (Morris 1948, 1956, 1964). Therefore, semioethics presupposes the human capacity for critique. Its special vocation is to evidence sign networks where it seemed that there were none, therefore connections and implications from which escape is impossible where instead there only seemed to be net separations, boundaries and distances with relative alibis. Such alibis serve to safeguard responsibility understood in a limited sense and, therefore, consciousness in the form of good consciousness, clean conscience.

Semioethics in semiotics, then, is a matter of fuelling biocentric human responsibility with theoretical practice.

Semioethics, in its cultural implications rather than in its absent programme, may still have a role to play in the public sphere. The term 'public sphere' is obviously invoked with reference to Habermas' (1989) groundbreaking thesis in

The Structural Transformation of the Public Sphere. Divorced from domestic and business concerns, the public sphere was, for Habermas, political in quite a 'pure' sense, a 'rational' discourse, not simply dictated by, or the epiphenomenon of, the accumulation of capital. One might consider, for example, how biosemiotics might have a relatively straightforward role to play in environmental considerations; Hoffmeyer's (2001) sketch of a semiotic view of bioengineering points to potentially revolutionary implications for farming. But, more generally, semioethics – or at least the biosemiotic premises that recast ethics – promise to come to the fore in an era of acute awareness of signs, an awareness which, indeed, faces its major threat not in a refusal to see the working of signs but in a failure to recognize semiosis in anything other than *human* signs. Even if many humans cannot recognize their own semioses as imbricated, commingled, continuous and co-terminous with non-human semioses, the implication of biosemiotics' understanding of ethics is that humans need to overcome the 'metaphysics of wickedness' that was identified through Peirce in Chap. 4, above. In the midst of the global financial crisis and, especially, in the wake of the banking scandals of the last decade, there is, much more than ethics in general, a fundamental need for the continuity of Peirce's thought-life to be a part of common sense. Perhaps the spate of subprime lending in the West at the turn of the twenty-first century will go down in history as one of the classic denials of the understanding that all thought is connected. Peirce (MS 1476, c. 1904: 38 – emphasis in the original), in concord with biosemiotics, indicates how such denial can inform erroneous individualist concepts of selfhood, for the "*entire thought-life of a person* and *the thought-life of a social group* and *entire body of all thought*" are all signs, albeit of much different types.

Undoubtedly, it would be a futile enterprise to try to persuade human groups to give up their pursuit of the core values associated with the good life. A voluntarist ethics repeatedly falls into this trap. It runs the risk of re-enacting a nature/culture division in which the human somehow has to be 'drawn back to' nature, forced into communion with *zoē*, leaving behind some of the scant comforts that 'culture' has afforded. The project of a biosemiotic ethics, as has been seen, is not like this. As Sebeok repeatedly pointed out, the idea of a split between nature and culture is absurd, simply because the latter is such a small compartment of the latter. Furthermore, if culture is in nature, then so, too, are its most exalted phenomena. Ethics is usually taken to be a 'cultural issue' while nature, of course, is red in tooth and claw. However, as the work of de Waal (de Waal 1996, 2001; de Waal et al. 2006) and others has suggested, there can be a translation of terms from the world of human ethics to the ethical behaviour of other animals. Because this ethical behaviour is not simply part of the pursuit of survival, a special kind of agency is involved. Yet, as has been seen, the analogous agency in human ethics has too often been presented as bearing a distinct complexion – discourse. This discursive character necessarily denies ethics to those organisms without language and limits the ways in which ethics can relate to non-discursive entities. Ethical imperatives towards the planet as a whole may have the tendency to consider non-human phenomena as either 'resources' for humans or as anthropomorphized (for examples, animals' eligibility for the equivalent of 'human rights'). Again, the insistence on discourse as the refracting lens through which humans 'see' is a kind of refusal of the cultural realm to allow translation from the natural realm as though both realms were completely unconnected with each other.

Even while de Waal's work remains ongoing towards its conclusions, there are further implications for culture – and nature beyond culture – in the biosemiotic understanding of ethics. There are even examples of practice inspired by it. For instance, the 'wooded meadow' discussed by Kull et al. (2003); that is, a place where biodiversity is preserved rather than threatened by human activity. Unlike the *wilderness* (where cultivation and human management are almost wholly absent) or the *countryside* (where nature is 'artificially' protected by humans), the Estonian wooded meadow is, to a small extent, managed, but includes mainly local (i.e. non-introduced) species (2003: 77). Care spills over into management, here, which represents a problem, in the long run, in recognizing the boundary between commitment and compulsion from the other. The wooded meadow initiative is spurred by a will, albeit one that is different from the will which would simply exploit the space.

Yet, in considering the dilemmas of ethics as normative, ethics as voluntary and ethics as ultimately unavoidable in the long run, there is one almost universal feature of animal life which possesses an intrinsic biosemiotic ethics, both as a displacement/projection activity and as the compulsion from the other. The universal feature in question is parenthood: a state of existence in an *Umwelt* where anticipation is very much to the fore. As well as this, parenthood is a state in which the well-being of sign collections continuous with but not identical to one's own is uppermost. Parenthood throws the self into relief through acute awareness of an other. Above all, the ethical imperative of parenthood seizes (in Badiou's sense, once more) the subject and demands fidelity without the subject ever knowing that 'ethics' are being enacted or that they are carrying out a programme. Notwithstanding cases of parental neglect in the worlds of human and non-human animals, as well as the fact that the growth of awareness of risk over the last 20 years in the human world has not conclusively fostered a more ethical social formation, parenthood arguably best exemplifies the biosemiotic conception of ethics. Parenting has been associated with myths of 'mother earth' in both the Western and the Eastern traditions (see Han 2016). In the semiotic tradition, in fact, the individualist and gendered bias of such clichés has been largely shorn to produce the adumbrated ideas of 'mother sense' (Welby 2009a, b) and 'Gaia' (Lovelock 2000). Parenting necessitates care without the kind of self-interest that Condillac and other humanists have assumed is in ethics. It is a semiosis that, for humans, contains experience of pleasure, pain, sadness and happiness in begetting further semioses of experience of pleasure, pain, sadness and happiness. It answers the call of the other. More than any of these, it is central to reproduction of the *Umwelten* of all species. Sometimes it is called love.

Chapter 6
Codes and Interpretation in Nature and Culture

Although biosemiotics stresses continuity of semiosis, it is also concerned with systematic patterning in semiosis, the points where recognition gives rise to meaning, the appearance of invariants, habit formation and the exercise of constraints. Its main concern is with interpretation, in which the agency of the organism in some way progresses semiosis; but, as discussed in Chap. 4, above, certain digital codes in the biosphere, identified in the concept of 'code-duality' do produce invariance and are important in biosemiotics. Because the idea of 'codes' is so prevalent in culture, cultural analysis and demotic speech, as well as having a part in the history of general semiotics, the current chapter will interrogate the kind of code notion that biosemiotics entertains in relation to interpretation.

Tracing the history of code as a concept or a phenomenon, rather than a practice, is not a an unproblematic endeavour. Most accounts of code are not self-reflexive. They seldom define code, nor do they give a sense of their history. In textbooks on mathematics, computing, information theory and cryptography, the discussion of codes begins with entropy and algorithms, the status of codes and their ontology (see, for example, Welsh 1988; McEliece 2002). Some mathematics textbooks allocate space to defining the concept of code; hence, Biggs (2008: v) states that coding is

> replacing symbolic information, such as a sequence of bits or a message written in a natural language, by another message using (possibly) different symbols. There are three main reasons for doing this: Economy (data compression), Reliability (correction of errors), and Security (cryptography).

Yet this seems a rarity. Most accounts of code go straight to cryptography in the Western tradition. The ancient Greek development of the *scytale* – a strip of paper with a written message that can only be read when wrapped in a spiral round a stick – and the Roman development of the Caesar code – a basic system of letter substitution in written messages – are often cited. It is telling that these examples come from a very advanced period in evolution, when humans had already developed writing as a form of non-analogue communication (although most accounts of

P. Cobley, *Cultural Implications of Biosemiotics*, Biosemiotics 15,
DOI 10.1007/978-94-024-0858-4_6

the origin of writing as rooted in economic transactions mention its roots in analogic communication).

The notion of code really only developed, especially in cryptographic terms, in the nineteenth century. As dictionaries of etymology will reveal, *code* has fraternity with the Latin *codex*, a book of laws; it is also related to *codicil* – an amendment to a document or a kind of over-writing of parts of a document. Subsequently, *code* was co-opted for what Singh (1999), in his popular science book, calls 'The mechanisation of secrecy' – sending encrypted messages. The code as a law, then, is overlaid with the sender and receiver's knowledge of the *rules* for decoding and the potential interceptor's lack of knowledge of those rules. It existed before the nineteenth century – in early writing systems or even non-writing systems for accounting, such as the quipu of the Incas developed nearly 5,000 years ago (Ascher and Ascher 1997). The common use of the word 'code' is recent, arising thousands of years into the process of routine coding by humans. The study of coding as part of a discipline only emerged, as so many disciplines themselves emerged, in the last 150 years; it is concurrent with the rise of electronic coding (from the telegraph, onwards) and fused, especially, with the development of communications, media and cultural studies in the second, third and fourth decades of the twentieth century.

For semiotics, the idea of code comes mainly from communications, but also more dedicated linguistic study. The key name associated with code-orientated linguistics, particularly during the fashionable days of semiology is, of course, Saussure (see Chap. 2, above). Saussure's projection of 'semiology', "a science *which studies the role of signs as part of social life*" (1983: 15) entails a 'synchronic' interrogation of the very conditions upon which signs operate. Saussure emphasized the language system (*langue*) underlying sign use, the sum of differences that occur between linguistic signs, none of which can rely on a natural process of 'meaning' but, rather, consist of the 'values' generated by each other's arbitrary relation between sound pattern and concept. Effectively, the language system provided a 'code' for generating coded speech acts. Saussure's perspective inspired others in semiology and linguistics, such as Hjelmslev and Barthes, to figure complementary ways in which one sign repertoire transforms into another one, such as an 'expression plane' and 'content plane' of signs combining to form a new 'expression plane'. However, there is a fundamental problem with applying 'code' to the understanding of language in this way. As Deacon (2012b: 10) notes,

> A code does indeed involve an arbitrary mapping or correspondence relationship, but that is precisely why its reference is opaque and is the basis for encryption. A code is a mapping of a parallel set of sign tokens to a language, and typically a token-to-token mapping. So to describe language or any of its attributes, such as the basis for phonology, syntax, or semantics as a code, merely begs the question: what is the basis for this mapping relationship?

Code in language, then, does not simply rely on arbitrariness in linguistic signs. There is something more.

The revealing of a semiotic code, in language or elsewhere, was already more complicated than Saussurean formulations allowed. Eco elaborated on the matter in the influential volume translated into English in 1976 as *A Theory of Semiotics*. Eco begins his discussion of codes with the example of an engineer in charge of a water gate between two mountains who needs to know when the water level behind the gate is becoming dangerously high. The engineer places a buoy in the watershed; when the water rises to danger level, this activates a transmitter which emits an electrical signal through a channel which reaches a receiver downriver; the receiver then converts the signal into a readable message for a destination apparatus. This allows Eco (1976: 36–7) to demonstrate that, under the designation 'code', the engineer has four different phenomena to consider:

(a) A set of *signals* ruled by combinatory laws (bearing in mind that these laws are not naturally or determinately connected to states of water – the engineer could use such laws to send signals down the channel to express passion to a lover);
(b) A set of states (of the water); these could have been conveyed by almost any kind of signal provided they reach the destination in a form which becomes intelligible;
(c) A set of behavioural responses at the destination (these can be independent of how (a) and (b) are composed);
(d) A rule coupling some items from the (a) system with some from (b) and (c) (this rule establishes that an array of specific signals refers to specific states of water or, put another way, a syntactic arrangement refers to a semantic configuration; alternatively, it may be the case that the array of signals corresponds to a specific response without the need to explicitly consider the semantic configuration.

For Eco, only the rule in (d) can really be called a code. However, he points out that combinatory principles that feature in (a), (b) and (c) are often taken for codes. This is consistently the case when such phrases as 'the legal code', 'code of practice', 'behavioural code' are in such wide circulation. Yet, what Eco's semiotics makes clear for the study of communication is that 'code' should strictly be taken as a 'holistic' phenomenon in which a rule binds not just the sign-vehicle to the object to which it refers but also binds it to any response that might arise irrespective of the reference to the object becoming explicit. At most, (a), (b) and (c) are to be taken as '*s*-codes' – systems or 'structures' that subsist independently of any communicative purpose. They can be studied by information theory but they only command attention from communication science when they exist within a communicative rule or code, (d) (Eco 1976: 38–46).

The other basis of Eco's theory of codes for semiotics concerns the character of codes (and *s*-codes) in interaction or 'sign-functions'. Rather than the 'referential function' by which a sign refers in a more or less direct way to an object in the world, in a synthesis of Saussurean and perspectives Eco stresses the way in which signs refer to other signs or 'cultural units': "*Every attempt to establish what the referent of a sign is forces us to define the referent in terms of an abstract entity which moreover is only a cultural convention*" (1976: 66; emphasis in the original). Thus, the 'meaning' of a term for Eco can only ever be a 'cultural unit' (1976: 67) or, at most, a psychological one. Moreover, this movement from one sign or cultural unit to another entails that signs are seen to work, along with communication, in a chain, a phenomenon that Eco discusses in terms of Peirce's conception of the *interpretant* (1976: 68–72). Significantly, Eco casts semiotics as a "substitute for cultural anthropology" (1976: 27), bequeathing to semiotics a model of communication which is orientated to the vicissitudes of culture rather than fixated on the possibility of referentiality.

Nevertheless, Eco's 'cultural' perspective is important to the consideration of the implications of the way that biosemiotics formulates code in relation to nature. As will be seen, particularly germane is the identification of different strengths of coding. When interviewed towards the end of his life, Eco interestingly observed that he and colleagues in the 1960s had been "incontinent" with regard to their touting of the strength of codes (see Kull and Velmezova 2016). Moreover, these strengths are likely to vary according to domain: the term code has one prevalent meaning and some more general ones – as Hoffmeyer (2008a: 83) notes, "the term has quite different connotations in different disciplines (e.g., jurisprudence, genetics, computing)". As will be discussed, the prevalent idea of 'code' as very strong and determining is sometimes taken for granted in a way which is responsible for elision between strong and weak coding. Yet, tracing the pivotal moment in the fate of the conception of code to Sebeok's development of zoosemiotics after 1963 will allow a tracking of the conception in biosemiotics which reveals notable implications for culture.

At present, in contemporary semiotics, it is clear that there are some leading definitions of code which, without some historical perspective, seem to be at odds. In his final book, Sebeok (2001b) repeatedly made reference to the five major codes: the immune code, the genetic code, the metabolic code, the neural code and, of course, the verbal code. These are all strong codes which need to be precise and mechanical in order for species to reproduce and survive. Hoffmeyer (2008a: 84–5) refers to code in general as the "customary use of distinct entities or actions for communicative ends" giving examples of a dress-code, a behavioural code, etc. These are weaker codes, more like guidelines, with some degree of flexibility. He adds that semiotics sees code in the frame of context-free rules for encoding, transmission and decoding as posited centrally to information theory and in a looser version of code as a vehicle for creation of meaningful activity, a semiotic resource that enables the creation and expression of certain types of meaning but not others. Barbieri (2003, 2010), another biosemiotician who has a broad take on the issue, sees semiosis in the genetic code, at the very origin of life, with context-free codes only being superseded by the development of interpretation by organisms 3 billion years later. Semiosis and interpretation, he adds, "are distinct processes" (2009: 239). This chapter will return to the latter point. In order to demonstrate how biosemiotics has a history of considering the questions of code, this chapter will first consider the evolution of the use of the term in Sebeok's work. Those who might challenge the cultural implications that arise from biosemiotics' formulation of code should see that the usual questions of code have been asked and a deeper consideration of code has also been carried out by biosemiotics.

Sebeok's use of the term code evolved in a recognizable way during his first zoosemiotic period represented by the 1972 collection of Sebeok's essays,

Perspectives in Zoosemiotics, as opposed to the second zoosemiotic period. Maran (2010: 318) identifies the first zoosemiotic period as 1962–1969, although this could be put back to 1960 when Sebeok delivered his first zoosemiotic paper at Burg Wartenstein; the second zoosemiotic period of Sebeok is identified by Maran as 1975 to the end of the 1970s; this could also be extended to encompass the 1980 New York Academy of Science colloquium on 'Clever Hans', the subsequent writings on the Clever Hans phenomenon in the 1980s and 1990s, and the essay on signifying behaviour in the domestic cat, published in 1994 but eventually included in Sebeok (2001b). What exemplifies the two zoosemiotic periods in Sebeok's writing is the place of code – it dissipates in the second period but, as will be seen in the focus on the first period, it is here in the relatively strong guise, in thrall not to information theory, in fact, but to an informationally-inflected linguistics (and, even so, already beginning the process of dissipation to the weak form).

The essays in *Perspectives in Zoosemiotics* are, to invert a cliché, an unmixed bunch. As Sebeok admits in the Foreword (1972: 4), some themes are repeated and some issues are re-stated for the different purposes to which the essays were originally geared. Nevertheless, it does seem that there was development of the notion of code in relation to animal communication as Sebeok continued to write throughout the 1960s. By reading closely and reproducing a number of quotes, it is hoped that that development is made manifest. The initial formulations on code offered by 'Animal communication' (published 1965) are fairly straightforward:

> A code is that set of transformation rules whereby messages are converted from one representation to another, a message being a string generated by an application of a set of such rules, or an ordered selection from an agreed, that is, conventional, set of signs (1972: 75).

The traditional association of 'code' and 'rule' is invoked, as is 'transformation' – possibly, but not definitively, a nod to Chomskyan generative grammar and its influence during the period. In an earlier paper, published in 1962 but dating back to 1960, Sebeok states (1972: 9) that

> By coding is meant an operation, governed by strict and logical rules, aimed at gaining increased efficiency by having elementary signals organized into patterns of alternative actions; by code is meant everything that the source and the receiver know *a priori* about the message.

'Rules' are evident here again, but Sebeok also has an explicit communication/information theory frame in the reference to 'source' and 'receiver'. This is not just a matter of context – Jakobson, one of Sebeok's key teachers (see Cobley et al. 2011), had formulated 'context' in his celebrated communication model of 1960 as a matter of the "referent", but he formulated 'code' as "common to the addresser and addressee" (Jakobson 1960: 353). Echoing the latter, the key phrase in the quote from Sebeok is "*a priori*" – existent and not depending on the situation. Thus, in the first part of 'Semiotics and ethology', which appeared in German, Czech and Polish in 1966, Sebeok (1972: 123) states that "The source and the destination are therefore said to fully, or at least partially, share (d) a code, which may be defined as that set of transformation rules whereby messages can be converted from one representation to another". This, then, is the general orientation of Sebeok's description of code in his early writings in zoosemiotics.

Yet, if the delineation of code as an *a priori* rule is scrutinized further, it is clear that Sebeok's 'code' has specific sources. In his 1960 paper, Sebeok (1972: 17) says of the human, "However, as to the code, his freedom is restricted: his selection must be made out of 'prefabricated units', among sets of simultaneous binary distinctions (Halle 1957), elements in an algorithm (or decision procedure) concatenated into sequential patterns". One telling phrase in this quote is repeated in his 'Animal communication' essay of 1965:

> The communicating organism's selection of a message out of its species-consistent code – as well as the receiving organism's apprehension of it – proceeds either in accordance with a genetic program dictating an almost wholly prefabricated set of responses with reference to each animal's unique memory store which then determines the way in which the genetic program is read (1972: 72).

The same phrase recurs a year later in the 'Semiotics and ethology' paper:

> In each species, the source of a message must share with its destination a code, the critical element of their communicative commerce constituting a particularized version of the universal 'need-to-know'. Every emitting organism's selection of a message out of its species-consistent code, as well as the receiving organism's apprehension of it, proceed either in accordance with a 'closed' prefabricated set of responses, or with reference to the animal's unique memory store which then directs the way in which the genetically precoded portion of the total behavior is acted out (1972: 129).

The key term in these quotes, to complement "a priori" here, is "prefabricated", as in the famous post-World War II housing, made in standardized form before construction and habitation. Code, as such, precedes communication.

The three sources for Sebeok's assertions about code are, in light of his central role in propelling international study of communication in the 1950s, predictable. They are (first order) cybernetics, communication theory and biology. A typical indicator of the influence of cybernetics can be seen in his 1964 'Discussion of communication processes' where he states that "the coding of information in cybernetic control processes and the consequences that are imposed by this categorization where living animals function as input/output linking devices in a biological version of the traditional information-theory circuit with a transcoder added" (1972: 84). This is clear cut. Slightly less clear cut, but typical of communication (and information) theory of the period if one is sufficiently acquainted with it, is the statement from the same 1964 source on

> the ultimate units of language, the atomic particles, if you will, of linguistic structure, organized in accordance with a binary code of utmost efficiency, so that whatever phase of the speech event is approached, the elicitation of its correlates must yield a distinct, unambiguous, 'yes' or 'no response (1972: 86)

Also unsurprising for those with a basic knowledge of Sebeok's career and intellectual trajectory is the inflection towards biology in his statements on code. Yet, in the 'Foreword' to *Perspectives in Zoosemiotics* it is significant that he states that "I became a professional linguist and, alas forever, a geneticist manqué" (1972: 2). It is well known that Sebeok repeatedly defined himself as "a biologist *manqué*" (e.g. 1991c: 9; 2011: 457), so defining himself as a *geneticist manqué* in this one place is significant. Perhaps it suggests that the theme of coding that is so evident in this collection of essays is, ultimately, allied with the putative invariability of genetics. In the 1968 essay on 'Goals and limitations of the study of animal communication', Sebeok writes that "It is amply clear even now that the genetic code must be regarded as the most fundamental of all semiotic networks and therefore as the prototype for all other signaling systems used by animals including man" (1972: 117).

Yet, in spite of the prominence of cybernetics, communication theory and biology in the discussion of code which runs through these early zoosemiotics essays, there is also a further, over-arching factor. That factor is linguistics, in particular Jakobsonian linguistics. *Perspectives on Zoosemiotics* is dedicated to the geneticist who taught Sebeok at Chicago, Joseph J. Schwab; but the figure who haunts its earlier pages is Roman Jakobson. In the 1964 'Discussion of communication processes', Sebeok gives an account of some of the key features constituting the human animal, concluding with the set of tools with which humans are endowed, identified by linguistics as "universal building blocks of language: these are called 'distinctive features'" (1972: 86). In fact, his account of distinctive features is clearly inflected by communication/information theory. Thus, the quote on this topic (1972: 86) given earlier can be reproduced here in expanded form:

> Distinctive features are the ultimate units of language, the atomic particles, if you will, of linguistic structure, organized in a binary code of utmost efficiency, so that whatever phase of the speech event is approached, the elicitation of its correlates must yield a distinct, unambiguous, 'yes' or 'no' response. What is basic and general in the structure of the expression in this form of human communication is, first, the minimal system of oppositions grounded on maximal distinctions; and, secondly, the rank order according to which this elementary phonological component pervades the more complicated syntactic and other constructions.

Sebeok, at this stage in his career, clearly considered distinctive features the "most concretely and substantively realized" (1972: 86) part of general linguistic theory, although he notes that the "phylogeny of distinctive features . . . has clearly not yet progressed beyond mere speculation" (1972: 88). To be clear, distinctive features are not the rules or code *per se*, but they are synonymous with the code to the extent that they constitute it. An analogous example might be football, where each individual kick or header or rebounding of the ball off the body is distinct but constitutes the code or rule by which handling the ball is outlawed. Jakobson had seen that the opposition of phonemes identified by Saussure (but, as Jakobson points out, introduced by the Polish linguist, Baudouin de Courtenay in 1870) was not a pure opposition, was susceptible of overlaps and was not sufficiently detailed to pronounce phonemes the basic, determinate units of language. Thus, in the 1940s and 1950s, Jakobson (Fant et al. 1952; Jakobson 1976) developed the theory of distinctive features in order to address the code underlying messages – that is, sounds more basic even than the phoneme; sounds that could not be reduced beyond their binary status.

In positing the fundamental constituents of the code underlying messages, Sebeok and Jakobson were effectively participating in the millennia-old practice of what Harris (1981) calls "the language myth". In an impressive body of work,

Harris's outlining of the key problem embodied in the language myth is purposely simple. It is that (in three words) "languages presuppose communication" (1981: 19); yet, in the face of this, the Western tradition of language study (as well as philosophy and then other areas of investigation, perfusing, ultimately, the laity and leading to a full-blown academic establishment called 'linguistics') has been intent on taking verbal communication apart, breaking it up into bits to attempt to determine what language is. This 'segregational view' "insists that 'the language' (the words, sentences, etc.) is one thing and what people do with it (or with them) another" (1996: 14). In the process of propounding a segregational view of language, successive 'administrators' of the Western tradition have taken as read that there are languages and parts of languages while, at best, only according secondary value to the fact that communication is a crucible of creativity in a swirling environment of changing contexts. Along with the larger myth, Harris identifies a number of sub-myths: the efficacy of dividing up speech to reach the 'essence of language'; sentences and propositions (especially in philosophy of language); language as facilitating telementation (especially espoused by Saussure); and the sub-myth that concerns the present article, the fixed code fallacy.

Jakobson veered towards a fixed code model of language chiefly through his rejection of Saussure's principle of linearity. Harris (2003: 96) points out that Jakobson chose an introductory passage on syntagms from Saussure's *Cours* and made the mistake of assuming that Saussure's principle of the impossibility of pronouncing two linguistic elements at once meant that two elements could not be *voiced* at once. The linearity principle seemed orientated towards context in that one element had to be processed in relation to another element; Jakobson's rejection of the principle led to his promoting distinctive features, sound distinctions at a level lower than the phoneme, which, as argued above, constitute a foundational, binary code. Jakobson's predilection for a theory of fixed codes in language was possibly a result of his interest in information theory in the 1950s, but it was also a logical product of 'segregational' linguistics. It is well known that Jakobson was very taken with the writings of Peirce from the 1950s onwards. Yet this does not guarantee a departure from code in favour of 'interpretation'. Peirce instituted his own version of code through the type/token distinction, as Harris (1996:10–12) recognizes. Jakobson focuses on this distinction, especially in the essays, 'Some questions of meaning' (1990a) and 'Quest for the essence of meaning' (1990b), but his linguistics appears to remain caught in segregational paraphernalia. In opposition to this, 'Integrational' perspectives propounded by Harris (see, for example, Harris 1998) and focused on *communication* not *language*, insist that communication is co-temporal and wholly context-dependent. This means that the moment of communication is the crucial matter and that that which is a priori in communication is limited and by no means consists of cardinal rules. Another way of stating the 'integrationist' view is to say that it proceeds from the idea that the message is not ineluctably dependent on the code. (Indeed, for all his work on distinctive features, there is evidence to suggest that Jakobson's reading of Saussure did not necessarily amount to a generativist account of code and message – see Jakobson 1990c [originally 1942]). While segregationists have dissected messages to discover a determinate form or a code in language, integrationists are more comfortable with indeterminacy of form in communication (see Harris 2006: 39–42).

It is as well for biosemiotics that the study of communication (as opposed to the study of language) has not been as slavish towards dividing messages and discovering codes. A central plank of animal communication studies for at least 50 years has been the insight that, when communicating, animals simply do not trade in the individual signs that might be demonstrated to be generated by a code. Rather, they send and receive 'whole' messages. One of the first ethologists to ruminate at length about this, Peter Marler (1961: 312), quoted by Sebeok in *Perspectives in Zoosemiotics*, wrote that

> in animal communication systems several items of information seem to be conveyed by one discrete, indivisible signal. We do not normally find the different items of information represented by different elements as is commonly the case in human language, where the component elements can be rearranged to create new 'messages'.

Since Marler wrote, this has become the accepted view in animal communication studies and has been borne out by much empirical research, such as that of Cheney and Seyfarth (1990; Seyfarth and Cheney 1993) into referentiality and specificity in vervet monkeys.

Certainly, Sebeok's later, 'fully semiotic' phase after the founding of the IASS in 1969 witnessed a different view on codes and coding from the context-free one. Looking back in 2001, he takes a jaundiced view of Birdwhistell's kinesics because the latter had "had minted this term by analogy with *linguistics* for the study of body motion from the point of view of how this may function as a communicative code" (2001b: xiii). In his later writings Sebeok referred to a proliferation of 'cultural' and 'natural' codes, from those in specific film genres to those in the social world of cats. He also treated the term 'code' as a synonym for 'interpretant' (see, for example, 2001b: 80 and 191 n. 13). Yet, even as the insights developed in zoosemiotics acted to question the theory of fixed codes in animal communication (including human communication), it is possible to see a change already during Sebeok's first zoosemiotic phase. In 'Zoosemiotic structures and social organization' (1972), Sebeok described different kinds of coding, (not unproblematically) considering Mozart's *Don Giovanni* to consist of a primary code – "natural language"; a secondary code – libretto; a tertiary code – score; and then the performance (1972: 164; a similar example appears in Sebeok (2001b). Sebeok's 'fully semiotic' phase and the dissipation of his conception of codes as fixed is perhaps indicated in his late comment that "Jakobson's impact on my linguistics studies having been pivotal – I should add at once that it was far less so on my gradual evolution as a semiotician" (2011: 459).

While the advent of zoosemiotics cannot be taken just as the cue for dismantling the notion of fixed codes – a dismantling that has thoroughly penetrated linguistics in the last 40 years but which was led by semiotics in general – it seems to be that in Sebeok's work and, by association, the institutional development of semiotics, zoosemiotics did have far-reaching consequences in respect of code. These are already prefigured in the 'first zoosemiotic phase' and can be whittled down to three related points that bear on semiotic research and theory: in 1968, Sebeok observed proscriptively that "descriptions of other sign systems tend to imitate (slavishly and erroneously) linguistics" (1972: 112); he noted that the "need for different kinds of theory at different levels of 'coding' appears to be a pressing task" (1972: 112); and, in 1972, he posed a key question for current investigation, "what is a sign, how does the environment and its turbulences impinge upon it, how did it come about?" (1972: 4). However, also in 1968, Sebeok remained convinced of the relation, as yet to be delineated, of the genetic and the verbal codes:

> the development of a normal neonate's faculty of language, which presumably includes a set of the universal primes of the verbal code, is wholly determined by the genetic code, but in such a way that this identical genetic blueprint can then find a variety of expressions in phenogeny through space and time (1972: 109).

Thus, it seems that in his research into the further reaches of the verbal code, the immune code, the metabolic code and the neural code, the importance of the genetic code's relation of determination was never absent.

In his final works before his death, Sebeok repeatedly invoked 'code' but he continued to oscillate between weak coding and strong coding. Thus, he uses 'code' to refer to

> Localized and conventional signaling systems – railway signals, smoke signals, semaphores, telegraph signals, Morse code signals, warning lights, flares, beacons, balefires, red flags, warning lights, traffic lights, alarms, distress signals, danger signals, whistles, sirens, bleepers, buzzers, knocking, gongs, bells, and drums (2001c: 10).
>
> communication in opera – Mozart's musical code, da Ponte's libretto plus additional nonverbal artistic codes, such as mime, scenery, setting, costuming, and lighting, among others (2001c: 16)
>
> film (3 codes), circus (5 codes), theatre (many codes) (2001c: 16)

In a glossary entry (2001c: 152), he defines code in an even looser fashion as a "system of signifying elements which can be deployed to represent types of phenomena in specific ways" and it is clear that he is being flexible to a point where explanatory efficacy becomes strained when he puts together an overview of codes:

> The genetic code, the metabolic code (hormone-mediated intercellular transactions), the nonverbal communicative codes used in a very high number of organisms including humans, our unique verbal code and its differentiated participation in all manner of artistic functions, whether literary, musical, pictorial, architectural, choreographic, theatrical, filmic, or of diverse hybrid formations, and finally, comparisons among any of the aforementioned – these all continue to be on the agenda of contemporary semiotic science (2001c: 114).

Sebeok does, however, retain the emphasis on those codes that have syntax: language and endosemiotic systems (2001c: 149), reaffirming in particular in the latter the genetic, neural, metabolic and immune codes (2001b: 72). Yet he suggests that "The question of an analogy between the two codes, the endosemiosic (molecular) and the anthroposemiosic (including a verbal component) seems, however, a secondary one. What matters is that both are productive semiosic systems." (2001b: 19). The elision, here, is not inconsistent with another late and curious statement by Sebeok on code. On the one hand, he states that a code is "a set of unambiguous rules whereby messages are convertible from one representation to another"; then, within the same sentence, he introduces ambiguity – or, at least, qualification – by asserting that a "code is what the two parties in the message exchange are supposed to have, in fact or by assumption, totally or in part, in common" (2001c: 31–2). Note: "supposed to have" and "in part". Then he offers an observation regarding the illusory nature of fixed codes:

> Using Joseph Weizenbaum's famous computer program, aptly named Eliza, human interlocutors tend to project sympathy, interest, and in telligence upon Eliza, as they would upon a psychotherapist. In fact, Eliza 'knows' nothing. A similar fallacy about shared codes is the theme of Jerzy Kosinski's brilliant novelette *Being There* (and the faithful movie based on it), in which an illiterate, retarded gardener is ascribed supreme gnostic attributes because he – essentially a blank page – mimics, echoes, and reflects back the interactive codes of every one of his conversational partners, whatever their native speech community may be (2001c: 31–2).

This statement seems to imply that the idea of fixed codes, as Harris revealed so energetically in respect of linguistics, does not really hold up. Nor can that idea really be considered credible in its rigid sense in the world of culture suffused by verbality. Indeed, as Pablé and Hutton (2015) have demonstrated, this is the key meeting point of contemporary semiotics and integrationism.

Yet, tarrying with fixed codes – or, at least, traversing the hinterland between strong and weak codes, as Sebeok did – is not unknown in semiotics. Because of the coding strength of 'syntactically-endowed' endosemiosis, the topic of coding has also provoked discussion in biosemiotics. In Chap. 4, the importance of 'code-duality' was noted, along with Hoffmeyer and Emmeche's (2007) qualifications. More recently, discussion on coding in biosemiotics has focused on the work of Marcello Barbieri, the theoretical biologist and co-founder of the international journal, *Biosemiotics*. Barbieri's project for a semantic biology developed in stages since the 1970s, but reached a particularly high point of accomplishment with his publication of *The Organic Codes* (2003), a monumental treatise written in an engaging style characteristic of fellow biosemioticians, Hoffmeyer, Deacon and, earlier, Sebeok. At the core of the volume is an elegant exposition of four principles and four models (2003: 244–53). These state that

> Epigenesis is a defining characteristic of life. Any living organism is a system that is capable of increasing its own complexity. What is crucial to life is not complexity as such, but the ability to produce a convergent increase of complexity. (First principle)
>
> Achieving a convergent increase of complexity is equivalent to reconstructing a structure from incomplete information. This implies a new definition of epigenesist as amounting to that task of reconstruction. (Second principle)
>
> The iterative algorithms that have been proposed for this reconstruction of structures from insufficient information actually perform two distinct reconstructions: one for structure and one for storage (memory). This takes place because organic epigenesis requires organic memories. (Third principle).
>
> There cannot be a convergent increase of complexity without codes. Organic epigenesis requires organic codes. (Fourth principle).
>
> "The cell is an epigenetic system made of three fundamental categories (genotype, ribotype and phenotype) which contains at least one organic memory (the genome) and one organic code (the genetic code)." (First model)

The second model of semantic biology, in conclusion, is the idea that "An animal is a trinitary system made of genotype, phylotype and phenotype." (Second model).

"Mental development is a sequence of two distinct processes of reconstruction from incomplete information, each of which increases the complexity of the system in a convergent way. The first process builds the specietypic mind (the universal grammar), while the second leads to the individual mind" (Third model).

"*The origin and the evolution of life took place by natural selection and by natural conventions. The great events of macroevolution have always been associated with the appearance of new organic codes*" (Fourth model; emphasis in the original).

Diachronically, these principles and models have produced the following lineage of life and code types:

4bn years ago: Origin of life – Genetic code
2bn years ago: Eukaryotes – splicing codes
2bn years ago: Multicellulars – adhesion codes
0.5bn to 0 years ago: Animals, Vertebrates, Amniotes, Mammals, Cultural evolution – pattern codes (Barbieri 2003: 233)

Along with this diachrony, there are some important corollaries contributing to the logic of succession of the organic codes:

1. The living forms which acquired a new organic code have never driven other forms to extinction.
2. A new organic code has never abolished previous codes.
3. The genetic code is present in all living creatures, but the other organic codes appeared in increasingly smaller groups, thus giving rise to a veritable "pyramid" of life.
4. Even if the evolution of an organic code could take an extremely long time, the "origin" of a complete code is a sudden event, and this means that the great evolutionary novelties associated with that code appeared suddenly in the history of life (2003: 234–6).

In the face of the length of this list of principles, models and corollaries, it should not be forgotten that the core of Barbieri's exposition is a major contribution to biosemiotics through the infusion of an awareness of meaning processes in nature: "The extraordinary thing about codes is that they require a new entity. In addition to energy and information they require meaning" (2003: 5).

Rather than regarding meaning as a spiritual or a transcendental entity, Barbieri suggests that it is "an object which is related to another object via a code". In its focus on code, this contrasts with other biosemiotic definitions of meaning which rely on wider processes such as 'recognition'. In a semiotic formulation indebted to the Saussurean tradition, Barbieri states that

the meaning of the word apple, for example, is the mental object of the fruit which is associated to the mental object of that word by the code of the English language (needless to say, the code of another language would associate a different mental object to the same word (2003: 5).

To this, Barbieri adds further examples of codes: dots and dashes represent a letter of the alphabet in the Morse code, a combination of three nucleotides represents an amino acid in the genetic code.

The account of codes in general in the volume embraces interpretation but not as 'choice' – interpretation is a process into which endosemiotic components are interpellated. In referring to signals' impact on the cell, Barbieri states (2003: 109) that experimental results have proved that

> outside signals do not have instructive effects. Cells use them to interpret the world, not to yield to it. Such a conclusion amounts to saying that signal transduction is based on organic codes, and this is in fact the only plausible explanation of the data, but of course we would also like a direct proof. As we have seen, the signature of an organic code is the presence of adaptors, and the molecules of signal transduction have indeed the typical characteristics of adaptors.

Usually, the fact that cells use signals to interpret the world does not necessarily lead to the conclusion that a code is in evidence, unless by a 'code' is meant weak coding. If it were strong coding that was in place, outside signals *would* have instructive effects. Later, Barbieri (2003: 182) also notes that the piece of the amino acid chain named the peptide leader that emerges first from the ribosome machine can contain a sequence that the cell interprets as an export signal to the endoplasmic reticulum. Only if the signal is not present, he explains, does the cell ignore it

That Barbieri reduces the level of endosemiotic interpretance to zero levels, raising the strength of the organic codes, is evident in the nevertheless groundbreaking passage (2003: 229–30) where he notes that modern biology has persisted with only two codes in nature: the genetic code, appearing with the origin of life and human codes of cultural evolution, which arrived almost 4 billion years later. Into that breach, as has been seen, Barbieri posits the organic codes. He states that

> the origin of an organic code is the appearance of a complete set of rules, because when that happens something totally new appears in nature, something that did not exist before. In the case of the genetic code, for example, we have already seen that its rules could have appeared one at a time in precellular systems, because each of them could give a contribution to the development of those systems. When a complete set of rules appeared, however, it was something totally new on Earth: what came into being was biological specificity, the most fundamental of life's properties. That event marked the origin of exact replication, the birth of the first true cells, and it is proper therefore to say that the origin of life coincided with the appearance of biological specificity, i.e. with the 'origin' of the genetic code (2003: 230–2).

Although Barbieri's principles and models demonstrate considerable congruence with the biosemiotic project in general, this statement leaves little room for some of the more familiar themes of agency, freedom, choice, habit, constraint and scaffolding that have been discussed in this volume.

Sebeok (2001b: 68) contended that "Because there can be no semiosis without interpretability – surely life's cardinal propensity – semiosis presupposes the axiomatic identity of the semiosphere with the biosphere". In light of what has been discussed in this chapter regarding Sebeok's movement to and fro between strong and weak coding, it is not possible to assume that "interpretability" refers to recognition of coded instructions or agentive choice. It is not surprising, then, that Kull (2012: 18) notes that the question of the relation of semiosis and code has been difficult to resolve in discussions among biosemioticians. He goes on to define code as

> a regular correspondence or link between entities that would not form such a regular correspondence on the basis of self-assembly (because, in cases where we have a code, there is an immense number of possibilities to form alternative links). As different from self-

assembly, the creating or inheriting of codes requires work; i.e., a code is a correspondence or link that is created or inherited by semiosis (by life) (Kull 2012: 18).

Kull concludes that "semiosis has primacy before codes; codes are products of semiosis". A code can persist after semiosis – in machines. He also adds that it is like "a frozen pragmatic" or "a frozen habit" (2012: 18). But it is semiosis that makes code-relations, their persistence, reconstruction and inheritance possible. Interestingly, Kull also argues that semiosis cannot exist without codes, that they are a necessary but not sufficient condition for semiosis. This, clearly, is a definition of codes as 'weak'.

What is perhaps the most persuasive part of Kull's argument concerns synechism, once more. Semiosis, he writes, "always requires a previous semiosis (*omne semiosis ex semiosis*; *omne vivum ex vivo* – except at their initial emergence at the origin of life)" (2012: 18). Something must come before coding – indeed, coding depends for its existence on that something: semiosis. That very semiosis accumulates the meaning that Barbieri suggests is needed for codes to work through recognition, memory, categorization, mimicry, learning and communication. Meaning is required for the functioning of codes, Barbieri attests; so it seems reasonable to suggest that code is not responsible for meaning-making but semiosis is: through the abutting and juxtaposing of more than one (weak) code (Kull 2012: 19). When a new code arises from semiosis, then semiosis has acted in a learning capacity, says Kull. One could add that it is also a recognition capacity. In either case, though, there is always the possibility of uncertainty or indeterminacy, a fact that distinguishes semiosis from code. Perhaps what happens with a code, because of its reliance on semiosic events that are not mechanical in their effect, is more like the action of an invariant or a habit, neither of which are 100% efficient. What is observed as a code in nature may be an anthropomorphism, imputing more cryptographic efficiency to an invariant than it warrants. Barbieri considers this, but argues that it "does not mean that a correspondence between two independent worlds must be the result of a conscious activity" (2003: 5). It is possible that the observation of the correspondence is still an anthropomorphism that extreme behaviourism would have avoided by simply noting that "this happens and then this happens". What matters, perhaps, and as was suggested in Chap. 3, above, is how the anthropomorphism is negotiated to reach the most revealing outcome or, in this case, whether the correspondences that are identified are deemed to be weakly or strongly coded.

Such work is synonymous with the term, 'interpretation'. To 'interpret' in English means both to explain the meaning of something and to translate the sense of something. Any event in the natural world that is observed by some entity that is *other than* that event will be deriving meaning from it and translating that meaning through the process of recognition, memory, categorization, mimicry, learning and communication. If the event observed is itself an event involving interpretation, then the matter becomes that much more complicated. Indeed, when the principles of interpretation are not finely tuned – perhaps they are overly mechanical or overly flexible – then interpretation becomes very fraught. Umberto Eco, who dedicated much of his academic career to this problem, fruitfully addressed these questions with reference to the occult philosophy and alchemy of Hermetism in the Renaissance. Among Hermetism's multifarious views was the idea that every symbol was related to like symbol, continuously. For example, some Hermetists thought that the plant orchis had a form which very much resembled human testicles (from the Greek orkhis = testicles). Therefore, every operation undertaken on the plant which gets a result would also get a result if undertaken on the human. By way of response to this potentially painful scenario, Eco describes the objection of Francis Bacon that one must distinguish between a relationship of causality and a relationship of similarity. That is, the roots of the orchis may look similar to testicles but the reason underlying the form of each is different. Eco (1990: 29), comparing this to Peirce's idea of habit as an alternative way to interpret says that, on hearing Bacon's objection Peirce

> would have added that, if the interpretation of the roots of orchis as testicles does not produce a practical habit allowing the interpreters to operate successfully according to that interpretation, the process of semiosis has failed. In the same sense, one is entitled to try the most daring abductions, but if an abduction is not legitimated by further practical tests, the hypothesis cannot be entertained any longer.

Two observations in respect of the topics discussed in this chapter follow on from this. The first is that properly strong codes would need to be completely mechanical and not open to interpretation in the sense of producing anything other than the intended result. The second is that strong codes can only be identified by way of abductions which run the risk of being fallible in the way that strong codes cannot. A further irony regarding strong and weak codes is revealed by the idea of code duality as raised in Chap. 4. Digital codes (strong ones putatively characterised by mechanism e.g. the notion of the 'selfish gene') have been imputed with an autonomous character when, in fact, their sphere of efficacy is limited. Conversely, analogue coding has acquired a reputation for independence and individualism when, in fact, individuals are subject to species history, mortality and the need to sexually reproduce (Hoffmeyer and Emmeche 2007: 51).

The implication for culture is one that *does* need to be spelled out. Cultural analysis, while knowing this implication, is prone to forget it. Not all coding is like cryptography, either from the encoding end or the decoding end. Nor is coding in culture identical to the digital codes of nature, although there are two features of the latter that are also often forgotten, that are most important and do recur in culture. The first is the syntactical bearing of endosemiotic digital codes such as the genetic code; the second is the fallibility of such codes as the genetic code which regularly effects mutation, the permanent alteration of the nucleotide sequence of the genome of an organism, rather than complete, reliable, uninterrupted reproduction of identity. Furthermore, the fallibility in recognition, memory, categorization, mimicry, learning and communication throughout nature is continuous with the recognition, memory, categorization, mimicry, learning and communication that occurs in culture. It should be remembered, of course, that there is much success in the operation of recognition, memory, categorization, mimicry, learning and communication – both in nature and its human compartment, culture. If there was not, then survival would be in jeopardy. However, the point here is that while semiosis involves codes, semiosis does not *amount to* codes. The invariants in semiosis can be made to work for the possible ends of *Umwelt* functioning and enhancement. The vicissitudes of those invariants, figured in what seem to be habits, constraints and even repression, are seldom visible in an obvious way. This issue is the focus of the chapter that follows.

Chapter 7
Freedom, Repression and Constraints

Semiosis has a tendency to grow, to lead to more semiosis. Yet, organisms often need to decelerate that growth, or repeat parts of it; this occurs through the appearance of invariants. As Peirce argues, it is the "essential function of a sign to render inefficient relations efficient – not to set them into action, but to establish a habit or general rule whereby they will act …" (8.332). Beyond invariance, an important issue in respect of the continuity of nature as it encompasses culture is that of the apparent impediment or blockage to straightforward development of a phenomenon. As will be seen in Chap. 8, the issue is occasionally overlooked in understandings of culture; invariably, though, there will be plenty of evidence to reveal that one or another cultural phenomenon has not had a smooth trajectory delivering it to its current stage of development. Instead, it will have been subject to overdetermination and uneven development. In Chap. 8, it will be shown that the descriptions of nature offered by biosemiotics need to be alive to overdetermination and unevenness, too. In the present chapter, the focus is on the conceptualisation of impediments to development, some of their consequences and how they are played out in relation to one aspect of culture in particular, the interface of the visual and the nonverbal more generally.

It would be foolish to imagine that evolutionary biology is impervious to multiple causes, despite popular accounts reducing it to key phrases. In Chap. 3 of *On the Origin of Species*, Darwin writes of the "struggle for existence", noting that variations, if they are in any way profitable to the individuals of a species "in their infinitely complex relations to other organic beings and to their physical conditions of life" (Darwin 1872: 49) will tend to preserve those individuals and be inherited by the offspring, giving them a better chance of surviving. This principle is called "natural selection, in order to mark its relation to man's power of selection" (Darwin 1872: 69). In the following chapter of *Origin*, in which he gives an extended definition of '"natural selection"', Darwin writes,

> Let it also be borne in mind how infinitely complex and close-fitting are the mutual relations of all organic beings to each other and to their physical conditions of life; and consequently what infinitely varied diversities of structure might be of use to each being under changing

P. Cobley, *Cultural Implications of Biosemiotics*, Biosemiotics 15,
DOI 10.1007/978-94-024-0858-4_7

> conditions of life. Can it then be thought improbable, seeing that variations useful to man have undoubtedly occurred, that other variations useful in some way to each being in the great and complex battle of life, should occur in the course of many successive generations? (Darwin 1872: 62–63)

Darwin is drawing attention, here, to factors of overdetermination. The key point is the complexity of relations in niches or enclaves of organisms and the "physical conditions of life". Of course, this stress is evidence that Darwin gave due consideration to the manifold nature of the conditions in which organisms exist, in contrast to the popular conception of natural selection as an immutable law. Nevertheless, there is a privileging of the conception of "use" in this statement, enforcing an unquestioned elision from "use" to "survival". Later, Darwin does try to mitigate that elision when he refers to the bee sting bringing the creature's own death, to huge numbers of drones being slaughtered by their sterile sisters, the waste of pollen by fir trees and *Ichneumonidae* feeding within the living body of caterpillars. He concludes that "[t]he wonder indeed, is, on the theory of natural selection, that more cases of the want of absolute perfection have not been detected" (Darwin 1872: 415).

On the issue of "use", these statements of Darwin remain mere qualifications. Famously, "use" has been challenged by Gould and Lewontin (1979) and Gould and Vrba (1982), in articles dedicated to unravelling the important distinction between current utility and historical genesis in evolution. More forceful for its deviation from the mainstream of evolutionary biology, however, is Hoffmeyer and Kull's (2003: 269) thesis that "use", in the sense in which it is envisaged in an "ecological niche", is superseded by "use" in the "semiotic niche" where the organism may have more control because it is in a conjunction in which all latently relevant cues have to be correctly interpreted. In a niche an organism does not attend solely to food, comfort and reproduction; instead, it must attend to an array of *signs* that are associated with those desirable entities (as it will also have to consider the signs of undesirable or "neutral", ignorable entities). Given the range of individual circumstances and signs that accrue, Hoffmeyer and Kull posit a Baldwinian perspective rather than a strictly Darwinian one since "organisms do not passively succumb to the severity of environmental judgment. Instead, they perceive, interpret, and act in the environment in ways that creatively and unpredictable change the whole setting for selection and evolution" (Hoffmeyer and Kull 2003: 269–270).

This not only extends the overdetermination that Darwin identified, but also unties the straitjacket constraining the complexity of niches in the Darwinian account. Still, overdetermination of niches has received only a partial theoretical framing. While Darwin binds survival with the principle of natural selection, Hoffmeyer and Kull free it through the agency which a semiotic environment facilitates. But, notwithstanding the latter's institution of a more agentive organism, neither consider the *quality* of survival. Nor is there a theorising of the nature of, on the one hand, the partiality in accounts of natural selection (Gould, Lewontin and Vrba are in a position to develop this, but do not) and, on the other, circumstantial limits on agency. Put another way, they do not theorise how "unfavourable" conditions might contribute to survival of organisms conceived as possessing agency. It seems that Darwinism, has not theorized how some things come to fruition and some things are prevented from coming to fruition. To put it yet another way, there is no account of the mechanisms by which, out of two favourable outcomes for survival, one might be 'repressed'. Nor is there a sense in which agency might precipitate one kind of survival rather than another. It is only recently, in fact, that biosemiotics has fine tuned such a theory.

Darwin had nothing to say on this issue, but another of the "great modern thinkers" whose work is, likewise, by no means unimpeachable, did. In his 1915 paper on repression, Freud states that an instinctual impulse may meet with resistances which seek to render it inoperative. If the impulse comes from an external location it can be countered by flight; but this alternative is not possible with an instinct, so the ultimate resistance, for humans at any rate, is *condemnation* based on judgment. The preliminary stage of this process of condemnation is repression, "something between flight and condemnation" (Freud 1984[1915]: 145). Yet, since satisfaction of an instinct is usually pleasurable, it is difficult to account for the internal stifling of an instinct or the transformation of it into unpleasure. Freud suggests that repression is therefore a matter of competing impulses in which the one that is repressed is, in fact, turned away or kept from consciousness (Freud 1984 [1915]: 147). Irrespective of whether one accepts the entire Freudian cartography of consciousness, his outlining of the terms of repression is nonetheless persuasive. Freud suggests that there is "primal repression", a first phase in which the "psychical representation" of the instinct is denied. This is followed by "repression proper", affecting mental derivatives of the repressed representative. Furthermore, the derivatives of the representative are said to each have their own vicissitudes. However, each case of repression is potentially subject to *displacement* and/or *condensation*. In the latter, the repressed idea is a receptacle for multiple causes beyond itself. In the former case, the repressed instinct is merely located to another idea or object—Freud gives the case of an animal phobia (the famous "Wolf Man", in fact) where repressed feelings in respect of the patient's father are worked out in relation to fear of wolves. Ultimately, Freud is no more able to say what the derivatives are and what determines specific repressions any more than Darwin is able to give a definitive account of "use". Both have to concede that specific instances are invariably massively overdetermined.

What is clear, however, is that the act of repression, like anxiety, is semiotic in nature: it contains an "idea" and associations to that idea. The instinctual impulse—about which Freud is sketchy—cannot really be conceived without its semiotic accoutrements. This may be one reason why psychoanalysis, despite having relatively little impact on psychology, has had some contribution to make to sociocultural thought, implicitly or explicitly. In semiotics, for example, a central—but largely implicit—concept in the work of Ponzio and Petrilli (Petrilli 2005; Petrilli and Ponzio 2005; Ponzio 1993, 2006a, b; Petrilli and Ponzio 1998; Petrilli 2014) is that capitalism, and latterly global communication, has constituted a sustained repression of dialogue, a force blocking the ultimate inescapable demand of the other. Typically, individualism has been the touchstone of this enterprise, but this has been accelerated in late capitalism through the promotion of monologic identity. In short: one set of impulses and associations advances while another is impeded.

Yet where some kind of repression is most evident in semiotic terms is in quotidian human interactions and their ontogenesis: the arena of psychology, perhaps, but, in place of the speculative sexual aetiology of psychoanalysis, semiotics offers a more reasoned evolutionary basis. Before considering impediments and blockages of development in nature at large, it is worth suggesting that the clearest example of a natural 'repression' with a major cultural consequence concerns nonverbal communication in human development. Although there has been little work carried out on this process, from the period when toddlers learn words, linking and elementary syntax there is a palpable repression of nonverbal communication. Indeed, this repression is institutionalised at this age by speech therapists and numerous "medical" tests to ensure that children are developing grammar. There has been opposition to, and struggle against, such linguistic imperialism—for example, by the deaf (Maher 1997). However, even this example leans towards a linguistic incarnation of communication while, in general, nonverbal communication in humans—gesture, proxemics, kinesics, music, visual communication of bodily changes—is viewed as a supplement to spoken language. It is this fact that makes Sebeok's (1988) (re) formulation of primary modelling all the more startling: as discussed in Chap. 3 (above), human "language" is shown to consist of an innate modelling device (geared to cognitive differentiation) which has become exapted for speech communication but, in the 2 million year history of the species (from *Homo habilis*), seemed to have allowed communication during the species' first 1.2 million years to take place via exclusively nonverbal means. Consider this matter, then, in relation to the 'visual' in culture – it offers an instructive case study in respect of the 'repression' or neglect of the nonverbal by cultural analysis and offers an opportunity to consider a new formulation of the process of 'repression' across nature.

As has been seen in preceding chapters semiotics is not only a discipline devoted to carrying out a micro-analysis of an artifact and then extrapolating from the findings some general observation about the artifact in question or the class of artifacts. From the early twentieth century onwards, including the spread of interest in Saussure's *Cours*, the benefits of close reading started to be enjoyed across the human sciences and semiotics was at the forefront of that. Yet, after biosemiotics especially, semiotics in its more contemporary guise consists not so much of micro-analysis, but in the act of stepping back to enable a broader view of how signification is organized in terms of media, modes, genres and species-specific semiosis. One manifestation of this contemporary programme has been the semiotic impulse to investigate 'the visual', rather than just 'visual artifacts'. Of course, something of the flavour of this has been offered outside of semiotics in respect to the impetus to identify 'visual culture' as a phenomenon characteristic of the contemporary social formation.

Developing in the 1990s, a number of commentators posited a pictorial, rather than 'textual', view of the world, where the 'world-as-text' was thought to be replaced by the 'world-as-picture'. In those heady days of the publishing venture known as 'postmodernism', many promoting 'picture theory' did so because they were identifying the new epoch as one which was dominated by the image (for example, Mirzoeff 1999; Mitchell 1994, *inter alia*). Possibly in an attempt to make this emphasis on the visual seem new, it was married with the masochism of French theory, particularly Foucault's *Discipline and Punish* (1977), to promote a refreshed version of what Martin Jay (1993) has identified as the "denigration of vision". Fredric Jameson's book, *Signatures of the Visible* (1990: 1–2) is exemplary in this respect: "The visual", he writes,

> is *essentially* pornographic, which is to say it has its end in rapt, mindless fascination; thinking about its attributes becomes an adjunct to that, if it is unwilling to betray its object; while the most austere films necessarily draw their energy from the attempt to repress their own excess (rather from the more thankless effort to discipline the viewer). Pornographic films are thus only the potentiation of films in general, which ask us to stare at the world as though it were a naked body.

This somewhat unequivocal assertion might come as a surprise to fans of *A Chump at Oxford* (1940) or *Rashomon* (1950), but it is a statement that is typical of the "ocularcentrism" that Jay sees in French thought and its epigones. Its masochism derives from the fact that it simultaneously delineates the walls of a prison from which there is no escape and nowhere to escape to—much like the narrative of the television series, *The Prisoner* (1967)—while longing for the outside world whose existence beyond its walls it has already denied. It matches the way Foucault's thought is torn between the omnipotence of all-pervading discursive formations and the longing for an anarcho-libertarian domain that the concept of discourse decrees is a figment of the imagination (Eagleton 2003; Levin 1997). The shadow of the 'Panopticon' (Bentham 1995; Foucault 1977) is constantly cast over 'visual culture' as a reminder of the supposed prison of almost total, controlling surveillance that humans have lived under since the Enlightenment. The 'visual' is seen as an implacable technology, its avatars mere versions of Michael, the camera lens-obsessed murderer in *Peeping Tom* (1960).

Contrast the position of 'visual culture', post-structuralism and ocularcentrism with that of the veteran neonate researcher, Daniel Stern. Recalling the birth of his interest in the ontogenesis of human communication, he writes (1998: 4)

> When I was two years old, I was hospitalized for many months for an operation that was complicated by an infection. In those days, antibiotics were not yet very effective and hospital stays could be quite lengthy. In addition, visiting for parents and family were fairly limited. At that age, I spoke only a few words and could understand very little of what was being said. But it was important for me to have a sense of what was happening. Like any child in that situation, I tuned into what people did, how they moved, what was happening on their faces and how they said what they said. In other words, I was paying attention to the music but not the lyrics, as these were beyond me. In short, I became a watcher and reader of the nonverbal. A lot depended on it.

Apart from the heart-rending nature of this account, in which the poor child is left to his own resources and proves most resourceful, modestly recalling later in respect of the nonverbal that "A lot depended on it", the quote indicates the first repression of nonverbal communication under discussion. The school of denigration of vision and its fellow travellers equate the visual predominantly with photographs and electronic media, entertaining the occasional foray into painting and other art-related practices. What they neglect—preposterously—is that visual technologies are just one minuscule portion of the entire sensory channel of sighted creatures. Of course, visual technologies can be argued to be extremely important as a crucial political battleground, particularly if they can be proven to influence or shape the way humans see. Yet, to forget that understanding the visual requires stepping back to examine how it functions for all sighted species effectively amounts to a repression of nonverbal communication. What Stern depicts in this quote is a world dominated by nonverbal communication, one in which vision is essential to survival and, tellingly, vision does the job adequately but by no means omnipotently.

Necessarily, 'visual culture' and the ideas that go with it—which seem quite old-fashioned now, although they have not vanished—are set up here as a 'straw man'. Nevertheless, they provide a contrast with a semiotics of vision, by which is not meant an apolitical, text-centred and self-centred hermeneutic, but an assessment of the role of vision in human semiosis. Thus, on the one side, there is a tradition and trajectory of thought in which discourse and the tyranny of vision is posited, along with an anarcho-libertarian hinterland which should not, according to that trajectory, exist. On the other side, there is a fast-developing trajectory in which vision is considered in terms of its embeddedness in an entire field available to the sensory channel of sight—a field which is highly diversified, features related forms and content and, through those widest of relations, offers the opportunity to determine how a species 'sees'. That field is not characterized by well-honed machines in a functionalist apparition of complementarity and control. Instead, it is riven with the potential for miscommunication and apprehends reality in only a fragmented way. It is a far cry from panopticism.

The field of nonverbal communication has been sufficiently variegated to have garnered some attention; and in recent decades, despite the historically and institutionally determined dominance of linguistics and the study of verbal communication, that field has even managed to fashion a place for itself in the academy (see, for example, Hall and Knapp 2013). In the popular imagination, nonverbal communication occupies a prominent position through the unfortunately designated 'body language'. The demotic understanding of 'body language', promoted since the 1970s in business manuals and popular guides (e.g. Fast 1970), is tacitly based on the notion that bodily communication among humans is highly codified and subject to a kind of 'grammar'. Sebeok (2001a) shows that this assumption is mistaken and argues that, like terminology such as 'the language of flowers', 'ape language' and so forth, the phrase 'body language' is to be avoided. When semioticians refer to nonverbal communication they are acknowledging the trafficking of signs *within* an organism or *between* two or more organisms (Sebeok 2001a). In humans, bodily communication comprises a number of elements. The most commonly recognized is manual communication or gesture (Kendon 2004). Yet there is also 'kinesics' (Birdwhistell 1970), made up of bodily movement and posture. As well, there is 'proxemics' (Hall 1966), focused on the orientation, proximity and distance of bodies as a matter of communication. These key features of human nonverbal communication, combined with general communicative attributes in the field of vision, have given rise to a number of media forms. These include mixed forms, such as theatre, with its combination of speech, nonverbal communication (bodily, and in set design/lighting etc.) and verbal communication. Film, television and, especially opera and other media that also incorporate music are supremely mixed forms (Sebeok 2001a). Yet, when making such observations, it is easy to forget that nonverbal communication inheres in the visible—or, as the ocularcentrists forgot, the visible is inherent in nonverbal communication—with respect to these media. The mise-en-scene of a film such as *Alien* (1979), featuring the justly famous interiors created by H. R. Giger, arguably carries out a large proportion of the communication in that narrative. The set design of a television soap opera such as *Eastenders* (1985-) no doubt communicates, visually, much of the vaunted 'realism' of that particular televisual form.

Yet finding this kind of integrated discussion of media, nonverbality and the field of the visual is not easy. One has to return to the classic, largely forgotten, text by Ruesch and Kees: *Nonverbal Communication: Notes on the Visual Perception of Human Relations* (1956). The authors set out their stall immediately, stating that

> the theoretical and systematic study of communication has serious limitations, inasmuch as scientific thinking and reporting are dependent upon verbal and digital language systems whereas human interaction, in contrast, is much more related to nonverbal systems of codification. Although most people are familiar with the rules that govern verbal communication—logic, syntax and grammar—few are aware of the principles that apply to nonverbal communication (1956: n. p.)

As they argue, much of the history of nonverbal communication has not been geared to the same kind of striving for representation that is characteristic of verbal and digital systems. As far as the visual arts were concerned, literal representation was hardly on the agenda before the Renaissance. Well into the Enlightenment, it was photography that provided the possibility, for the first time, of disseminating information at length nonverbally. Clearly, for Ruesch and Kees, the development of scientific thought on the back of writing and then printing in the Enlightenment has served to place further emphasis on the verbal/digital incarnation of knowledge, such that scientific knowledge of human communication has remained depressingly scant (1956: 12). Even with the putative increase of nonverbal semiosis in large amounts, from the arrival of the still photograph through moving pictures through Web 3.0, the idea that "culture is becoming more visual" (e.g. Ibrus 2014) would probably cut no ice with Ruesch and Kees. The problem they identify is also connected to the way that disciplines and subject areas develop in the academy.

What is known as 'the visual' has had a strange, but not uncommon, institutional predestination. As Machin (2014: 5) notes,

> where a new' realm of investigation is 'discovered' it can then herald a new flurry of activity that can, to those outside looking in, appear rather arbitrary. New network leaders will emerge in this new pioneering area of research. New terminologies will appear to account for the very same things already documented decades before in a different field ... In my own field of linguistics, the specialism of 'multimodality' has over the last decade seen linguists draw models from their own field to attempt to identify the building blocks of the visual: a visual grammar. But it soon became clear that these scholars were largely

ploughing the same furrow as over a century of semiotics, yet still not asking very basic and important questions about the nature of visual signs that had long been standard fare in this long standing tradition.

Elkins (2003; also cited by Machin 2014: 5–6) in particular has been outspoken about the spurious limitations placed on his field, noting the fixation on a clique of theorists and a constrained set of interests that do not proceed far beyond websites, some aspects of television and still photography. Yet, the shortcomings of the 'visual culture moment' in the academy remain a straw man because, as Machin rightly argues, they are actually symptomatic of the way many fields develop in the nexus of universities, publishing and higher education policy. The more serious problem is that the growth of new knowledge is stymied by disciplinary protectionism and the wilful neglect of holism. Machin (2014: 6, 9) adds,

> What really is the justification and use in analysing the visual apart from other modes of communication, from language, sound and materiality? Most of the communication we come across happens in different modes simultaneously … [I]n fact a wider view of visual communication is one which does not disconnect it from other modes of communication and is in fact the very study of human action and culture.

Effectively, what Machin calls for here is an approach and a field with the same disciplinary principles that Ruesch and Kees adumbrate in their classic work. One name for this would be 'semiotics of vision' or 'visual semiotics' or, at the very least, 'semiotics'.

How would such a 'visual semiotics' proceed? Well, the name would only be illustrative because it would seek to restore the connection of the visual to other modes of communication. Arguably, the first step should be backwards, to facilitate the kind of wider, species-aware view that has been promoted by biosemiotics. Sebeok thought in precisely such broad strokes and outlined the channels for signs or the channels in which communication takes place:

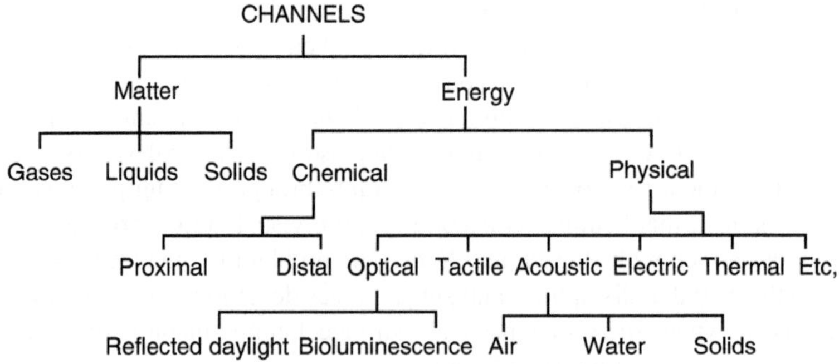

(Sebeok 1991c: 27)

The 'visual' can be found in the optical channel, a physical manifestation of energy which is facilitated by light. The other channels (tactile, acoustic, etc.) are facilitated by other phenomena in the universe. A visual semiotics would not only apprehend the location of the optical channel and its relations; it would also be compelled to ask about the Umwelt that was being invoked in the study of vision, acknowledging that humans are not the only sighted creatures and that the commonalities between the vision of humans and non-human animals need to be considered along with the differences. Hence, Sebeok also indicates the sources of signs:

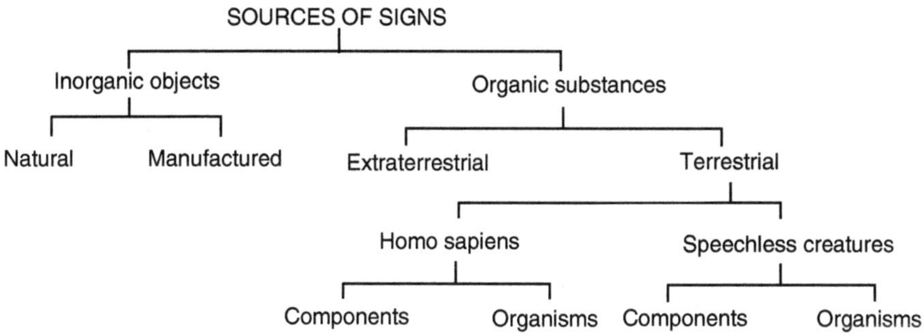

(Sebeok 1991c: 26)

The clear division here is, first, between organic substances and inorganic objects; then, second, between the speechless creatures and *Homo sapiens*. What unites the latter two, however, is that they communicate from organism to organism, but also within organisms. Visual semiotics, then, would be concerned with visual artifacts in an ecology of component/organism semiosis. Or, to pitch the matter in a more digestible prose, visual semiotics would look at how visual artifacts operate in association with the range of communication around them, as Machin proposes. This does not mean that the most fruitful work in 'pictorial semiotics' (Sonesson 1989) is to be abandoned; however, the *pars pro toto* fallacies of 'visual culture' and the like are to be regarded with suspicion.

Now, it is possible to return to the guiding issue of this chapter: that there is natural blockage, impediment or even repression in human development—in this case, in relation to nonverbal communication where it relies on the optical channel. What is under discussion here is a *phylogenetic* and *ontogenetic* matter. As observed in Chap. 3, above, and onwards, the human *Umwelt* can be understood as being derived from an innate 'modelling' device by which humans can differentiate the world. Humans use their sensorium in a comprehensive fashion, utilising, too, the zoosemiotic nonverbal and the anthroposemiotic verbal (Sebeok 1988). As has been seen in previous chapters, this means that there are sign systems (nonverbal communication) which, in terms of evolution, are antecedent to, and give rise to, externalised linguistic sign systems. Nonverbal communication is recognised by Sebeok as an adaptive communicational capacity possessed by all living beings. So, in the development of this modelling, something must be lost with the movement to one mode from another. In the theory of natural selection, it is clear that what gets lost are the species, or species members, who do not adapt fit features to the evolving environmental imperatives. Biosemiotics, on the other hand, has been critical of the ruthless mechanism of the theory of natural selection. Contra neo-Darwinism, it posits ‘semiotic freedom’ and elements of learning in evolution. For example, Hoffmeyer refers to experiments where scientists placed artificial sweeteners rather than glucose in the environment of a chemotactic bacteria cell. He writes (2010b: 164),

> In such cases, it seems appropriate to say that the cell misinterprets the chemical signs of its environment. Such misinterpretations are dangerous, and natural selection will favor any solution that helps the organism to better interpret the situations it meets. Indeed, selection would be expected to favor the evolution of more sophisticated forms of “semiotic freedom” in the sense of an increased capacity for responding to a variety of signs through the formation of (locally) ‘meaningful’ interpretants. Semiotic freedom (or interpretance) allows a system to ‘read’ many sorts of ‘cues’ in the surroundings, and this would normally have beneficial effects on fitness. Thus, from the modest beginnings we saw in chemotactic bacteria the semiotic freedom of organic systems would have tended to increase, and although it has not been easy to prove that any systematic increase in complexity, as this concept has traditionally been defined, has in fact accompanied the evolutionary process, it is quite obvious that semiotic complexity or freedom has indeed attained higher levels in later stages, advanced species of birds and mammals in general being semiotically much more sophisticated than less advanced species.

This semiotic freedom characterizes the scaffolding process in evolution, where the organism ‘builds’ on its relation to the environment. Hoffmeyer’s further development of the concept, generalizing it to cover the network of semiotic interactions connecting an organism with its *Umwelt*, shows how it facilitates processes of perception and action. The piecing together of parts of scaffolding, as has been seen in Chap. 4, above, produces particular reproducible ‘meaning’ for an organism as it takes part in the functional cycle of *receiving* signs appropriate to the sensorium and *producing/circulating* sensorium-appropriate signs. As Hoffmeyer (2010b: 164) explains, the process of scaffolding, traversed by semiotic freedom, contains something akin to a ‘goal’:

> Allowing for semiotic freedom in the organic world significantly changes the task of explaining emergent evolution, because semiotic freedom has a self-amplifying dynamic. Communicative patterns in assemblies of cells or individuals may often have first appeared as a simple result of the trial-and-error process of normal interaction, and may then endure for considerable periods of time. If such patterns are advantageous to the populations (cells or organisms), they may eventually become scaffolded by later mutational events. Through this ‘semi-Baldwinian’ mechanism, the evolutionary process will enter a formerly forbidden area of goal-directedness.

Thus, the semiotic freedom of organisms is responsible for its survival, for its evolution and contributes to changes in its environment.

Yet, such descriptions, in presenting a functional process, often run the risk of overlooking possible impediments or by-products of forward-looking mechanics. What about those occasions when one choice is made over another? Something has to be lost or left behind. Sometimes what is left behind is something that it is beneficial to lose, such as negative memories (Ritchie et al. 2015); sometimes leaving something behind has deleterious consequences. The matter has to be considered in biosemiotics because, apart from anything else, it is part of agentive action. Semiotic freedom necessarily involves choice of one course rather than another. In studying such freedom, there is often a need to investigate the choices that get rejected (and why), particularly as they may later become choices once more or there may be opportunities for the organism to revisit or relive the moment of choice. In the case of the phylogenetic development of communication, it is clear that the 'choice'—exaptation—of linear speech for human communication was significant. By no means did it eclipse nonverbal communication; nor did it demote nonverbal communication to a subsidiary role in real terms; but it did ensure a bias towards the verbal and a disregard for the nonverbal that effectively banished such communication to a realm that is not conscious in the way that it was for earlier hominid ancestors.

A related fate can be seen with respect to ontogenetic repression of human nonverbality. In infancy, the child is almost solely reliant on nonverbal signs. Its *Umwelt* is attuned to verbal signs and such signs will certainly circulate there; but those same kinds of signs will not yet emanate from the child her/himself. For the infant, as Stern (above) suggests, a lot depends on nonverbal communication. Around 18 months, however, the child with an expected development rate will start to use speech and syntax in an elementary fashion. It is for this reason that children's development is usually tested at that time: in Europe, this principally takes place through the public health system. The results of such tests may enable a decision to make an early intervention in those cases where the child is not developing as expected, indicating, through this symptom, auditory or cognitive problems. Such tests administered at 18 months in the UK are geared to literacy, grammar and syntax. The child's powers of concentration are observed, while the main focus is on the child's ability to understand words and, above all, link them in sentences. Yet, the following are *not* tested or observed: skills in drawing, gesturing, singing, sense of body space, rhythm, powers of mimicry, etc. The unpredictable nature of young children's behavior and attention will mean that at least one of these skills will invariably manifest itself even in the controlled circumstances of the test. Yet, such skills are not the focus of the test or taken as indicative of cognitive potential.

That infant innate powers of nonverbal communication do not simply disappear from 18 months onwards is powerfully affirmed by the work of stage magicians. In sleight-of-hand, *legerdemain* and *prestidigitazione*, they pull off tricks that, by virtue of their seemingly occult mechanisms, amaze onlookers. Yet, such tricks are almost totally dominated by mastery on the part of the magician, and forgetting on the part of the audience, of 'lost' nonverbal arts. Lions can therefore be tamed by the re-learning of the niceties of proxemics. A coin can be surreptitiously pocketed at the moment a seemingly insignificant gesture distracts the onlooker. The magician can predict the answers of an audience member, simply by 'muscle reading' their kinesics. The audience could have developed all of these skills possessed by the magician; but, without the magician's dedication and focus on the task, it would have taken a lifetime.

At a certain stage in phylogenesis, it is reasonable to assume, the field of human sight was more attuned to the nonverbal communication which took place in the human's environment. With the advent of linear speech, humans became creatures unique in their possession of verbal and nonverbal modes. Machin's lamentation, above, regarding the separation of modes for the purposes of academic study is therefore entirely apposite in this light. Fixation on the visual apparatus of some technologies alone is not only unhelpful, it is also impractical since so many 'technologies of vision' are already verbal-audio-visual. Furthermore, such fixation also represses the fact that the field of the visual is irrevocably embedded in the encompassing field of nonverbal communication in general (which also includes an ineffable number of transactions currently invisible to the human eye). Moreover, there is a predicament that subtends all of these matters: that is, the problem inherent in the phylogenetic and ontogenetic repression of nonverbal communication. That there has ever been an impetus to isolate the 'visual' and to treat it as a realm of (semi-)autonomous functioning—or, yet more problematic, dictated by linguistic principles rather than associated to them by dint of a common modelling ancestor—is symptomatic of a grave error. It is precisely that error which has given rise to the idea of 'visual culture' and the want of sobriety in the assertion that culture is becoming more visual.

Apart from its role in making communication more efficient in specific contexts and putatively contributing to the survival of the species, there is a further reason for the persistence of nonverbal communication in the face of its "superseding" by verbality. Aside from efficiency, nonverbality is frequently pleasurable. This is evident in that nonverbal communication plays such a prominent role in the performances of magicians, in music, in the feats of "intelligent" animals (e.g. Clever Hans), in the exercise of vision (Cobley 2011) and in rudimentary reasoning—abduction (Peirce 1929; Sebeok and Umiker-Sebeok 1980). Each involves a confrontation with repression and each, at least momentarily, involves an unblocking of the human potential that has been left as residual by the choice of the path of verbality. To be sure, repression is not simply a matter of blocking pleasure for the sake of it; much repression allows human communion to take place and society to be feasible—this is the position of the later Freud, in fact, in works such as *Civilization and Its Discontents* (Freud 1985[1930]), although his gloomy prognostications in this respect are not final. The matter at hand concerns repression of "bad", unpleasurable things, as well as "good", pleasurable ones. Some kind of repression has to be considered in biosemiotics because it is part of agentive action. Semiotic freedom necessarily involves choice of one course rather than another.

In biosemiotics it is important to consider whether this kind of repression operates at a level of semiotic freedom which has developed only in the human, or whether there is repression at the level of lower organisms. Certainly, the degree of semiotic freedom available to organisms is proportional to what is left over, courses that are not chosen, actions that do not come to fruition. Yet, the question remains regarding which organisms enact repression as described above. One clue might be offered by Peirce and one of Sebeok's footnotes. Sebeok (2001b: 96) asks whether the one-way ethological implication among the three categories of "taming/training/domestication" might be analogous to the Peircean categories of Firstness/Secondness/Thirdness and whether these map onto Charles Morris's programme for semiotics: syntactics/semantics/pragmatics. The suggestion in the present discussion—which turns out to be completely unoriginal—is that repression of instincts is required for animals to become domesticated. What is slightly more original is the idea that repression is somehow involved in Thirdness. Interpretants, as Thirdnesses, produce some new signs and not others; rhemes efface their origins in qualisigns; induction belies its abductional roots; the pragmatics of communication encourages some interpretations or perlocutions and not others. Thirdness, or the movement towards Thirdness, seems to harbour repression as an indigenous component. Yet, there is a need to be clear about what is an occlusion and what is a nesting. Indices, for example, are not really occluded; they are nested in symbols. Possibly, this is what is occurring with the nonverbal residues of communication in language.

There is a further big issue to consider in relation to biosemiotics and repression. This concerns the repression of "continuity" in the sense of the non-voluntary ethics described in Chap. 5, above. The process of parenting among animals involves "instinctual" protection of offspring exceeding, or on a par with, self-protection. The offspring exists, in this light, as an "other" but also as a pre-eminently valued extension of the self. When the bird leaves the nest or the carnivore's offspring goes off to hunt for him/herself, or when the human reaches adolescence, there arises the necessity to repress the extensional relation of the offspring to the self (although the most aged of human parents attest that the repression is never fully effected). Phylogenetically, the human's "consanguinity" with all organic life on earth seems to have undergone an analogous repression—necessary for carnivorous and herbivorous nourishment, perhaps, but, as domestic pets and horticulture attest, the repression is far from absolute. Ultimately, this point about recognizing continuity of the realms of life is one that Deely et al. (2005) seek to make, albeit clouded by a willed "ethical" programme (see Chap. 5, above). The general argument arises from the biosemiotic concept of semiotic freedom and moments when that freedom is constrained.

What semiosis often entails is not the efficiency of the strongest of coding which would lead to identical reproduction, as was discussed in Chap. 6, above. In fact, much of semiosis is concerned with local rather than global interpretations and can involve imperfect recognition, memory, categorization, mimicry, learning and communication. Far from proceeding with absolutely predictable results, seemingly mechanical processes betray their incomplete properties. Deacon (2012a: 104–7) gives the example of a computer that crashes because its current function is in a loop, engaged in an interminable machine process. The common response to this problem is for the user to turn the computer off and then on again. He notes (2012a: 104),

> If interference from outside the system (i.e., outside the mechanistic idealization that has been assigned a given computational interpretation) is capable of changing the very ground of computation, then computation cannot be a property that is intrinsic to anything.

Thus, it can be concluded that computation is an idealization about cognition based on an idealization about physical processes. The extrinsic, simplifying constraints—switching on and off—with respect to the computer's mechanical operations are determined in the context of operations that are prevented or otherwise not realized. As Deacon (2012a: 105) adds, "Paying attention to what is not occurring is the key to a way out of this conceptual prison". What establishes these constraints? The obvious answer is "the human brain"; but, as Deacon observes, this merely "passes the explanatory buck", only for it to get passed again to 'evolution' and 'natural selection'. What needs to be accepted is that real world physics and chemistry do not simply act mechanically like a computer with the occasional bug but, in fact, the kind of noisiness and messiness implied by a bug characterises their operations as a whole. This kind of imprecision was also observed in the last chapter with reference to codes. In Deacon's terms, the messiness is not just consonant with the second law of thermodynamics but is a matter of the 'ententionality' (incompleteness) with respect to which phenomena are organized for achievement of something that is not intrinsic to them.

The concept of constraints is of some considerable importance to Deacon's account of ententionality in nature because, as he concludes (2012a: 538), "Mind didn't exactly emerge from matter, but from constraints on matter". Leaving that somewhat large issue aside, the concept of constraint has implications both for biosemiotics and culture. As Deacon suggests (2012a: 191–2), constraint "is a complementary concept to order, habit, and organization because it determines a similarity class by exclusion". What is important here is that

> the concept of constraint does not treat organization as though it is something added to a process or to an ensemble of elements. It is not something over and above these constituents and their relationships to one another. And yet it neither demotes organization to mere descriptive status nor does it confuse organization with the specifics of the components and their particular singular relationships to one another. Constraints are what is not there but could have been, irrespective of whether this is registered by any act of observation (2012a: 192).

As Peirce held for habits, regularity or organization—rather than any specific substrate—is most relevant in respect of causation. "The term *constraint*", writes Deacon (2012a: 193), "thus denotes the property of being restricted or being less variable than possible, all other things being equal, and irrespective of why it is restricted".

As examples of constraints, Deacon offers the way a fast flowing stream forms stable eddies round a rock and how a snowflake grows hexagonally symmetric but idiosyncratic branches. In growing, the branches of the snowflake "progressively restrict where new growth can take place." (Deacon 2011); in this way, "Constraints reflect what is not there, and the more constrained something is, the more symmetric and regular it is" (2011). Already, this is a more nuanced picture of invariance than codes or even habits allow. It also seems to recast what has thus far been discussed in relation to repression. Rather than an 'information theory' version of 'constraint', what Deacon proposes is an invariant with a capacity for recreating its "capacity for self-creation", where "self" is not that far away from Sebeok's self (see Chap. 4, above) as "an intrinsic tendency to maintain a distinctive integrity against the ravages of increasing entropy as well as disturbances imposed by the surroundings" (Deacon 2011). Ultimately, Deacon posits dynamical reflexivity and constraint as characterising a teleodynamic system (2012a: 510). Yet the "constraint-preservation process" sheds light on the role of invariance in agency in general, for it is "the simplest exemplar of an intrinsically end-directed process, whose most fundamental end is maintenance of itself" (Deacon 2011). Constraints are not necessarily to be conflated with order, although the ideas are related; Deacon explains

> As in the case of the messiness of a room, order is commonly defined relative to the expectations and aesthetics of an observer. In contrast, constraint can be objectively and unambiguously assessed. That said, order and constraint are intrinsically related concepts. Irrespective of specific observer preferences, something will tend to be assessed as being more orderly if it reflects more constraint. We tend to describe things as more ordered if they are more predictable, more symmetric, more correlated, and thus more redundant in some features. To the extent that constraint is reduced variety, there will be more redundancy in attributes. This is the case with respect to any change: when some process is more constrained in some finite variety of values of its parameters or in the number of dimensions in which it can vary, its configurations, states, and paths of change will more often be 'near' previous ones in the space of possibilities, even if there is never exact repetition (2012a: 195).

It is for this reason that Deacon suggests that the concept of constraint could supplant habit

> Recasting the Realism/Nominalism debate in terms of dynamics and constraints eliminates the need to refer to both abstract generals, like organization, and simple particular objects or events lacking in organization. Both are simplifications due to our representation of things, not things in themselves. What exist are processes of change, constraints exhibited by those processes, and the statistical smoothing and the attractors (dynamical regularities that form due to self-organizing processes) that embody the options left by these constraints.

Or, as Deacon repeatedly states, the crucial issue concerns what is "not there".

In light of the previous discussion of repression in respect of nonverbal semiosis of humans, there is, perhaps, a need to revise the estimation of the action in question. "Repression" seems to suggest the smothering of some entity that needs to be free. It is an intimation that there is some anarcho-libertarian hinterland that might be reached if repression were lifted. If this is found to be unsatisfactory, an alternative explanatory principle needs to be critical in avoiding a mere functionalist bearing before consideration of its implications for culture can be undertaken. The idea of constraint seems to fit that alternative bill. On the one hand, it does appear to explain only 'successful' processes:

> And it is ultimately the production and propagation of constraints that make physical work possible. For example, containing the expanding gases in an internal combustion engine, and thus constraining expansion to occur in only one direction, allows this release of energy to be harnessed to do work on other systems, such as propelling the vehicle which contains the engine up a steep incline. So to argue that constraint is critical to causal explanation does not in any way advocate some mystical notion of causality (Deacon 2012a: 203).

Yet, on the other hand, to this statement is added a critical coda which accounts not just for what is there but the nature of the trade-off regarding what is lost: "We can restate this causal logic as follows: reduction of options for change in one process can lead to even greater reduction of options in another process that in some way depends on the first" (2012a: 203). As with the snowflake and the messy room, greater regularity has its costs.

There are, no doubt, many implications for culture arising from the way in which constraints generate redundancy. However, the next chapter will outline some of the consequences of not understanding the centrality of constraints to the culture compartment of nature. Culture is necessarily ententional and what is "not there" in culture may be the evolutionary point: culture's 'pointlessness'.

Chapter 8
Humanities Are Natural

There is a very real constraint on culture in the contemporary world, as opposed to the Deaconian constraint which is arguably at the centre of culture. The humanities are currently under assault for their perceived lack of utility. The humanities are found wanting in the face of the putative utility of science, technology, engineering, and mathematics (STEM), and they are increasingly called upon to demonstrate direct economic use-value. Subject areas such as medieval history are seen by critics of the humanities as being arcane, over-specialised and divorced from the brute economic realities which are supposedly paramount in contemporary life.

Without wishing to draw too facile a distinction, semiotics is accused of being over-generalised, despite having some of the flavour of practicality that is imputed to the sciences; the humanities are accused of being over-specialised and without demonstrable utility. Although the intent here is not to rely on this distinction, it does serve as a starting point to discuss the pratfalls of a knee-jerk defence of the humanities, and to suggest that a more nuanced response to the assault on liberal arts education in general – a response which might be decisively informed by biosemiotics – could be put centre stage in common understandings of what the humanities are for. That a more convincing response to the assault is desperately needed is demonstrated by the fact that the squeezing of the humanities, and the universities that house them, has accelerated even in the face of two key events in the last 15 years.

First, in the wake of 9/11 there was a commonly-held view that the terrorists used education in a purely instrumental fashion; *The 9/11 Commission Report* assiduously lists the university affiliations of the main conspirators, all of whom studied science and technology, apart from Hani Hanjour who sojourned in the United States to study English and later took flying lessons (Kean et al. 2004: 160–5). Indeed, some have pointed to the prevalence of ex-engineering students in terrorist attacks (Popper 2009; Gambetta and Herzog 2007), ultimately leading to the question "Is there something in an engineering education, such as that of 9/11 attacker Mohamed Atta, that, due to a lack of a component of humanities study, could lead to a lack of compassion for others?" (Bryson 2010). Second, the financial crisis of 2008 brought to the fore

P. Cobley, *Cultural Implications of Biosemiotics*, Biosemiotics 15,
DOI 10.1007/978-94-024-0858-4_8

much hand wringing that had been already fomenting in business schools (see Ghoshal 2005), centred on the dehumanizing process of business education. As the full extent of the catastrophe of subprime lending at the turn of the twenty-first century was becoming clear, many called for a renewal of the humanities and an infusion of liberal arts into business schools (for example, Colby et al. 2011).

Yet such considerations have cut no ice with governments. In the UK, for example, a key plank of the post-2010 Tory government's policy was to cut all funding to humanities in universities through raising fees for all humanities subjects. That the humanities as a whole is failing to articulate its worth in contributing to the activity of the mind in the current climate is cause for concern. Addressing this from the standpoint of biosemiotics, some of the cultural implications from previous chapters will be discerned, along with some instances where those implications have been totally overlooked – by instrumental policy makers, obviously, but also by those in the history of the humanities who have had the best interests of culture in their sights.

The 'rise' of the humanities can be traced back to Cicero's concept of *humanitas* – being good – and its development in Western education, particularly the *trivium* and *quadrivium* of medieval philosophy faculties, embracing humanities and natural sciences alike, as against the professions (medicine, law, theology). Closer to the present time, though, the humanities in their most familiar form are a product of nineteenth-century Western education: they developed in tandem with the forging of a liberal hegemony in industrial society of that period and contributed to the reproduction, through instruction – in what is civilized and 'good' – of the bourgeois class in their mercantile and civic incarnations. Again, the philosophical faculty contained humanities as well as sciences (as is still the case in the Liberal Arts programmes in the US), while the natural sciences only became autonomous in the latter half of the nineteenth century. The *decline* of the humanities has arguably occurred steadily through the same period in the face of the rise of the natural sciences (Kagan 2009), but most rapidly with Western governments' promotion of STEM in the academy during recent decades, managed through a crisis of funding.

As far as business schools have been concerned, the putative humanizing value of the humanities has been asserted repeatedly at crisis points in late capitalism. During the Cold War, McAllister's quasi-ethnographic study *Business Executives and the Humanities* (1951) gave voice to numerous managers who valued, above all, a liberal arts/humanities background for their recruits. These aspirations or requirements were echoed later in the decade by the Carnegie Foundation study (Pierson 1959) and the Ford Foundation study (Gordon and Howell 1959), each concerned with business and higher education. In the Reagan era, the American Assembly of Collegiate Schools of Business report (Porter and McKibbin 1988) made similar noises, followed in turn by the report of the American Council of Learned Societies (1988). By the early twenty-first century, a full-blown crisis in business schools seemed to have developed globally, with numerous critics calling for the re-humanization of business education, usually by way of compulsory humanities modules. Ghoshal (2005) has already been mentioned; preceding him, Pfeffer and Fong (2002) and Mintzberg (2004) could be added, along with, later,

Bennis and O'Toole (2005), Starkey and Tempest (2006), Starkey and Tiratsoo (2007) and Morsing and Rovira (2011) and those reporting inept practice by business school graduates (Feldman 2005; Blasco 2009). A recent milestone in this train of thought is the Carnegie Report, which concluded (Colby et al. 2011: 5): "Like all undergraduates, business students need the ability to grasp the pluralism in ways of thinking and acting that is so salient a characteristic of the contemporary world". That pluralism, which the report suggests is fostered by the humanities, is assumed to be lacking in business graduates but also, it might be said, among religious fundamentalists, particularly those who would inflict terror.

If the situation was not sufficiently overdetermined already, the last decade also saw a major crisis in Western universities as a whole. In another ethnography, lightly 'fictionalized', Tuchman and Wannabe (2009) pithily illustrated some of the nodal points of the crisis, witnessing the adversarial pitting of a management class against an intellectual class and the 'deprofessionalization' that has beset university professors in similar ways to its infliction on lawyers and doctors. With managers in the ascendant, along with context-free accountants scouring university spreadsheets (no doubt following an education in business that the authorities in the previous paragraph would deplore), it was unsurprising that questions began to be raised by apparent ingenues about what universities are for. In addition to asking whether it is really worth employing certain professors and buying certain equipment for universities, accountants' questions about the contribution of certain subject areas to direct economic growth become inevitable. As Collini (2012: 144–5) notes,

> [I]t's usually at this point in the argument that an appearance is made by one of the more bizarre and exotic products of the human imagination, a wholly fictive place called 'the real world'. This sumptuously improbable fantasy is quite unlike the actual world you and I live in. In the actual world that we're familiar with, there are all kinds of different people doing all kinds of different things – sometimes taking pleasure in their work, sometimes expressing themselves aesthetically, sometimes falling in love, sometimes telling themselves that if they didn't laugh they'd cry, sometimes wondering what it all means, and so on. But this invented entity called 'the real world' is inhabited exclusively by hard-faced robots who devote themselves single-mindedly to the task of making money. They work and then they die. Actually, in the fictional accounts of 'the real world' that I've read, they don't ever seem to mention dying, perhaps because they're afraid that if they did it might cause the robots to stop working for a bit and to start expressing themselves, falling in love, wondering what it all means, and so on, and once that happened, of course, 'the real world' wouldn't seem so special anymore, but would just be like the ordinary old world we're used to. Personally, I've never been able to take this so-called 'real world' very seriously. It's obviously the brainchild of cloistered businessmen, living in their ivory factories and out of touch with the kinds of things that matter to ordinary people like you and me. They should get out more.

He is not wrong. Indeed, Collini's characteristically witty observation should serve as the standard riposte to any blinkered imbecile who dares to hide behind the myth of the economically hard-nosed 'real world'. However, as will be argued, Collini's eloquent defence of the humanities as worthwhile amidst the university crisis – because they are "inherently" good or interesting – is not tenable on its own.

In response to the more recent attacks, the defence of the humanities has been undertaken by numerous of its representatives besides Collini in the last few years, often re-hashing jaded ideas from the very liberal hegemony which has lately sought to condemn the humanities to, at best, marginal status in society and, at worst, oblivion. Thus, the humanities have been cast by their defenders as the repository of 'values' (McDonald 2011) or, even more pointedly, 'good' values as opposed to "our current values and their devastating consequences on a precarious world" (O'Gorman 2011: 281). The humanities, it has been claimed, teach people how to live their lives (Andrews 1994: 163), they condense collective experience (Bate 2011: 66) and they preserve both democracy (Nussbaum 2010) and civilization (Watt 2011: 205). A further confection on liberal protestations in favour of saving the humanities is located at the intersection of national languages, ethics, and multiculturalism. Other languages, the argument goes, enrich our culture (Kelly 2011; Freeman 1994) and allow knowledge of 'the other' in a fashion that, at the very least, provides the platform for an ethical standpoint. The humanities are seen as crucial to promoting diversity – teaching students to work with others who are not like them (Tuchman 2009: 208) – because, unlike approaches in some business schools, for example, the humanities are putatively opposed, in their very existence, to de-humanization. Echoing psychologists such as Zimbardo and Milgram, as well as prominent critics of business education from *within* business schools, such as Ghoshal (2005) and De George (1994), Nussbaum (2010: 23) insists that "It is easier to treat people as objects to be manipulated if you have never learned any other way to see them". The acme of such humanist hyperbole regarding the humanities is where such arguments reveal their fragile basis and give way to the ridiculous; the words of the broadcaster and academic, Mary Beard (2011: 26), on the preservation of classics because it "is a subject at which the British do very well indeed", reflected by Parker Pearson (2011) on archaeology and Howard (2011) on British academia in general, lie in this domain.

By contrast, there is a sublime position growing out of the definition of the humanities as fostering harmony or standing against de-humanization. Here, the discussion of the immediate use-value of the humanities is repudiated in favour of a subtle formulation of inherent worth. Bate (2011) shows that the 'value' of the humanities cannot be calculated in the immediate way that many translations of scientific developments into technological advance can. In the wake of 9/11 and resurgent Islamic fundamentalism, he writes (2011: 2), "it was perhaps unfortunate that the swingeing funding cuts to higher education in the early 1980s fell with particular severity on supposedly marginal areas of the humanities, such as 'Islamic Studies'". More emphatic, still, is Fish's (2008: 14) refusal to rise to the challenge:

> To the question 'of what use are the humanities?', the only honest answer is none whatsoever. And it is an answer that brings honor to its subject. Justification, after all, confers value on an activity from a perspective outside its performance. An activity that cannot be justified is an activity that refuses to regard itself as instrumental to some larger good. The humanities are their own good. There is nothing more to say, and anything that is said … diminishes the object of its supposed praise.

Fish, here, is responding in particular to those who would attempt to furnish the humanities with 'effects' or 'results' in the manner of some areas of the sciences and business. Nevertheless, it is a view broadly shared with some other contemporary commentators (cf. Collini 2012) on the threatened demolition of universities.

There is a need to be clear about the terminus of such arguments about the humanities. Fish and others seem to be converging on that well-known shibboleth, common to discussions in many degree review and validation processes in universities: 'knowledge for its own sake'. While it is a worthy aspiration, it is strictly an intellectual version of the Land of Cockaigne, the preserve of those with private incomes. Likewise, the notion of the humanities as a civilizing tool, a less trenchant view but one nevertheless similarly drawing on individualist and humanist roots like 'knowledge for its own sake', is ultimately self-defeating. Tuchman (2009: 208) sees the humanities as promoting diversity and teaching students to work with others who are not like them; O'Brien (2010: ix) and Nussbaum (2010: 7) insist they are essential for democracy; and Pugès (2011: 61) claims they are instrumental in understanding other cultures and experiences, enabling people to keep an open mind. All of these arguments, however, are functionalist: they see the humanities as social tools, rather than necessary extensions of humans' cognitive bearing as a species. Thus, the obvious example of 'intercultural communication', a laudable area of investigation in communication sciences, was swiftly co-opted as a management tool, in much the same way as 'ethics' and 'diversity' are now (Nelson et al. 2012). It is difficult to escape the conclusion that "The humanities is an often overindulged and oversold commodity, especially in the hands of liberal arts college presidents and some recent secretaries of education" (Solomon 1994: 48). It is also clear that the version of the humanities that is oversold is not necessarily familiar to those who teach and publish in the discipline.

In his excoriation of business school practice, Ghoshal asks (2005: 83–84):

> why does the pessimistic model of people as purely self-interested beings still so dominate management-related theories? The answer lies not in evidence but in ideology ... The roots of the ideology lie in the philosophy of radical individualism articulated, among others, by Hume, Bentham, and Locke.

As is also argued in the current essay, Ghoshal is pointing out that if one wishes to address ideology – including that ideology which has culminated in an attempt to banish the humanities – then the last people one would want to consult are humanists. The project of de-humanization which is integral to the subordination of people to so-called the 'real world' is a logical outgrowth of the ideology in which humans are compelled to realise themselves as individuals – *at all costs*. Althusser (1969) made this point, in compelling fashion, many years ago. However, it has not curtailed the assumption, on the part of those outside the humanities, that the humanities is predicated on, and begets, both humanism and individualism.

Nevertheless, the simple point can be asserted: the humanities are not necessarily humanist. Indeed, the virtues that the humanists have found to be universal and enriching have been repeatedly repudiated as oppressive by such fields as postcolonialism. Communications, media and cultural studies have consistently, implicitly and explicitly, challenged such humanist edifices as the canon and authorship of the 'best', while introducing questions to do with the fragmentation of contemporary identity. Semiotics has done the same – but more systematically, with commitment to transdisciplinarity and without automatic disdain for science. When one considers such features of the modern humanities landscape which are not entrenched in a humanist liberal arts paradigm, then many other approaches and fields start to add their names: social constructionism, poststructuralism, deconstruction, posthumanism, systems theory, radical constructivism and, again, postcolonialism. In their most productive guises, what has characterised all of the above has been a commitment to transdisciplinarity.

Although semiotics traces its genealogy back to the Hippocratic Corpus of symptoms, its presence in the academy as a formal pursuit owes much to what might be called 'the synchronic moment' in the twentieth century (cf. Deely 2010). That moment, when analysis of the products of human endeavour gradually started to replace valorization of discrete cultural artifacts, was also key to the inauguration of transdisciplinarity. It was represented by the work of Saussure in Switzerland; Propp and the Formalists in Russia; Ogden, Richards, Empson and Leavis in Britain; the New Criticism, Innis, McLuhan and Frye in North America; the structuralists in France; the Prague Linguistic Circle in Czechoslovakia; so-called 'Soviet semiotics'; the Copenhagen School in Denmark; systems theory and cybernetics in Europe and the Americas. Thus, the synchronic moment, where close reading or analysis came to the fore, witnesses a significant change in some of the key disciplines of the humanities in the second part of the twentieth century. Linguistics became less concerned with teaching foreign languages and more dedicated to the workings of language in general, drawing, especially, on semiotics' separation of linguistics into 'syntactics', 'semantics' and 'pragmatics'. A good proportion of contemporary literary study became devoted to analysis of the workings of literariness rather than trying to wheedle out what is 'the best' that has been thought and written. An indication of how far literary studies has come through reinventing itself in the last 30 years is offered by some of the innovative work emanating from the Society for Literature, Science and the Arts. In the Fine Arts which, by virtue partly of their name, are still somewhat wedded to notions of the sublime, there has nevertheless been a move away from pure aesthetics to greater consideration of the concept of design. In philosophy, the elusive 'good life' has been superseded by a focus on analysis, criticality and unpredictability.

The prime mover in the majority of these instances has been the emergence of the idea of the *text*, developed, of course, by semiotics but with a remit and reach that has not only facilitated transdisciplinary approaches but also made the text's predicates part of common parlance in the humanities. In the early writings on the topic, by the idea's simultaneous but unconnected originators, Barthes (1977a) and Lotman (1982), it is possible to discern the struggle to make the concept emerge (cf. Marrone 2014). Despite this struggle, the testimony to text's fecundity is in how quickly the concept was taken up by other scholars in the wake of the synchronic moment. Such scholars were bearers of a transdisciplinary perspective on their subject areas, demonstrating by reference to 'text' how manifestations of art, literature, philosophy, and verbal language are not instances of magic but specific exemplifications of a more general textuality. Clearly, the notion of text was instrumental in closing the 'great divide' (Huyssen 1986) between high and low culture. Central to the movement away from appreciating 'quality' in cultural artefacts and a movement towards analysis of texts is the dimension of social class. The concept of the text betokens 'neutrality' or, at the very least, the attempt to shelve the ephemeral forces that may valorize or render a text in a particular way such that it is read in a fashion that is 'self-evident', 'common sense', or 'obvious' (cf. Cobley 2000). Therefore, the purpose of analysing a text is to find out how it works and, by extension, to help accumulate a sense of how all texts might work. Academic engagement with the text in this frame is decidedly *not* an exercise in 'distinction' (Bourdieu 1984) geared to the inculcation of good 'breeding' or 'taste', the inevitable product of upbringing in a well-financed bourgeois home. It is more of a technical skill potentially accessible to all, as befits a democratic society.

The transformation wrought by the concept of 'text', shifting the focus from the 'good' to the 'analytic', is the defining feature of contemporary humanities, although one would not know it if the only evidence on offer was that of the humanities' humanist defenders. Yet, while the insights attendant on the concept of 'text' suffuse the humanities, largely abolishing the hierarchy of 'high' and 'low', there is one deconstructed division arising from transdisciplinarity that the academy is more slow to accept: that between science and the humanities. While this is not the place to chart the fortunes of the 'two cultures', it is possible to briefly outline two areas where semiotics has contributed to the closing of the division. The first is relatively straightforward: it derives from the idea that if the humanities can be read as text, then there is absolutely no reason why the practices of science cannot be read as text also. Indeed, semiotics has given birth to one of the foremost exemplars of the understanding of nature with reference to textual and semiotic principles – the focus of the present volume: biosemiotics.

Second, there is the more complicated critique, at the level of philosophy of science, whereby the arts and humanities are placed alongside the sciences in a sometimes non-hierarchical relationship between different kinds of knowledge of the universe. Cybersemiotics (Brier 2008a), comprising much of biosemiotics, compels a vision of life, consciousness and cultural meaning as constituted by the continuities of nature and evolution. In this, it does not differ from general semiotics in the contemporary period; however, cybersemiotics specifically addresses life/consciousness/cultural meaning with reference to the qualities of experience each renders. Yet it does have a commitment, like general biosemiotics, to the process of knowing involved in an Σ-science (see Chap. 3, above). It challenges physicalist science, with its ideal of third person knowledge, replacing it with an imperative to consider first person embodied consciousness. Organism, environment, cognition, signs and reality – none of these are issues to be settled by one discipline. For this reason, cybersemiotics is transdisciplinary, tracking those areas in the humanities and the sciences where there have traditionally been materialist, organismic orientations in understanding phenomena and where there have been semiotic, cognitive orientations, also seemingly dictated by the phenomena with which they have been most concerned. This is summed up by Brier's (2010: 1907–11) "cybersemiotic star":

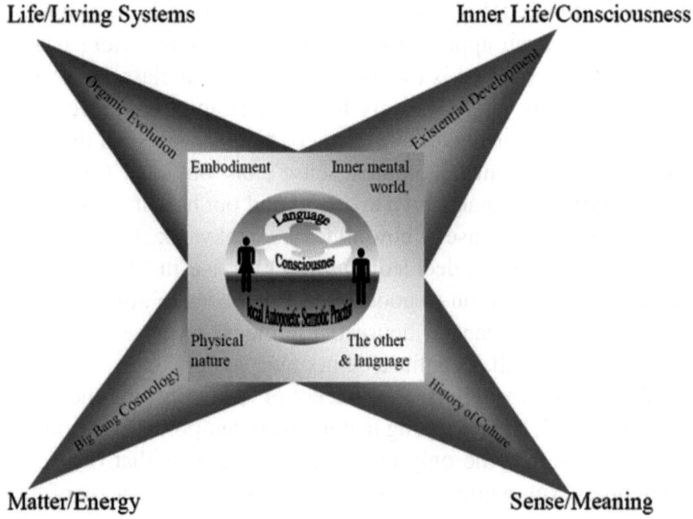

The four areas of knowledge that cybersemiotics identifies obviously demand transdisciplinarity. Moreover, as Brier argues, they also demand a theory of the observer. Physics, he notes (2010: 1911), relies on the notion of an observer of physical 'events' but

> it does not have a theory of what the observer is that goes further than computation and information … Meaning, experience, qualia and will are still outside that paradigmatic foundation of physics which, through chemistry, leads into general cell and body physiology

One might easily add to this that the humanities need a theory of the observer. The 'other' humanities, particularly through semiotics, have been diligent in questioning the role of agents in the world. The humanists, on the other hand, have seemed to fall back on the assumption of an absolute, universal human agency, even as that agency is being nullified in the dismantling of the humanities. For this reason, the convincing articulation of what the humanities are for depends on a stance that is anti-humanist.

The humanism that has often been taken as synonymous with the humanities can be summed up as "in short, bringing out what is best in us" (Solomon 1994: 50). This ideology is clearly evident in so many of the protestations against the assault on the humanities that have been quoted so far. One can understand the knee-jerk response: Churchwell for example (2014: 29), is strident:

> The politicians and corporations telling us that the humanities do not matter are, by no coincidence, the same people who think of us only as workers and consumers, not as citizens or individuals, and who strip away our human rights, one by one. It is the wealthy who insist that we should seek only to work: we don't need the humanities, they tell us, all we need is to labour in the marketplace that will enrich them, not us.

What is left out here is that 'they' very much believe in the individual; it is precisely why 'they' want to limit the opportunities open to others. The only collectivity 'they' can envisage without fear is the one that 'they' seek to impose. Clearly, the fundamental terms of the argument need to change from their individualistic/humanistic co-ordinates that are so tied up with the right to self-determination and enrichment. From the side of the humanities, such arguments are of a piece with the idea that the task of the humanities is to exalt "The best that has been thought and said". However much the proponents of humanist humanities may think they have left such views behind by teaching about women writers, black artists, Navajo verbal expression, and Lao Tzu, their defence of the humanities in terms of breeding and the 'good' resuscitates the ghosts of Matthew Arnold, F. R. Leavis, Mortimer Adler, Robert Hutchins and Lionel Trilling. In putting a notion of 'the human' at the centre of existence, the plight of the guardians of the 'good' is the "theoretical unevenness" that Althusser (1969: 223) discerned in 'socialist humanism'. Amidst the legacy of the terror and totalitarianism of the Soviet Union, many Marxists (including inside the Soviet Union) found themselves in a dilemma in denouncing this most prominent embodiment of supposed socialism. Althusser (1969: 236) suggests that socialist humanists fall back on a simplistic couplet, 'human/inhuman':

> When, in the relations between Marxists and everyone else, the former lay stress on a socialist personal humanism, they are simply demonstrating their will to bridge the gap that separates them from possible allies, and they are simply anticipating the movement trusting to future history the task of providing the old words with a new content.

With 'man' as the principle of all theory (see Chap. 4, above), a shadowy concept of 'inhumanity' acting somewhere as 'man's' obverse is needed by humanism. In this way, humanism can have some purchase as a practical, ideological slogan, rooting out instances of 'inhumanity'. Humanism, additionally, may have some value as a "practical index" (1969: 247) – in the case of humanist protestations against cuts in the humanities, perhaps as 'propaganda' – but it is only "an imaginary treatment of real problems" (ibid. 247); it has no theoretical value.

Thus, anti-humanist thinkers in biosemiotics – such as Peirce, Sebeok, Hoffmeyer, Deacon, Brier, Petrilli and Deely – do not put the individualized human at the centre of existence. Nor do they trade in essences such as 'self-interest' or apply universal categories to people. They certainly do not take the tack of the arch-humanist, Condillac, in formulating *ethics* as a matter of self-interest (see Chap. 5, above). The anti-humanism in biosemiotics, in particular, envisages humans within semiosis and within *Umwelten*. Human agency is not a matter of standing outside semiosis and administering signs like an air traffic controller, as humanist understandings of the humanities would have it. Human agency *is* the Umwelt; humans are *within* the products of semiosis that make up the objects of the humanities.

One could say, broadly, from a semiotic perspective, that there are two kinds of agency. The first might be called 'sociosemiotic', deriving from humans' situation vis-à-vis semiosis in cultural formations (see Cobley 2014). The other kind of agency is 'biosemiotic', in the realm of semiosis which is putatively not subject to the vagaries of cultural or socio-political forces (see, also, Chap. 4, above). This distinction, of course, is problematic for three main reasons. First, all semiosis is 'social' in character in that it involves more parties than just one (Cobley and

Randviir 2009); secondly, culture, as Sebeok noted and as has been repeatedly emphasized in the current volume, is just one small compartment of nature; thirdly, as Agamben (1998) and others attest, semiosis 'in nature' is more frequently subject to the vagaries of socio-political forces than is often acknowledged. Nevertheless, agency has become a central theme in biosemiotics (Tønnessen 2015) and is instructive for the question of what the humanities are for. Biosemiotics has identified agency at very lowly biological levels, in the most rudimentary of organisms. For Hoffmeyer (1998), it is possible to identify agency in any organism that develops 'semiotic competence' in the semiosphere – that is, in any realm in which signification or communication may take place. Biosemiotics has been at pains to demonstrate the occurrence of semiotic competence in places that have not hitherto been considered for their agency. As far as the humanities are concerned, this is an important point because it not only indicates some measure of continuity across some components of humans and other organisms, but it also suggests a crucial issue: the ways in which agency is 'inhabited' in an *Umwelt.* Of course, there is need for caution, because agency, as has been seen, is clearly taken for granted in the humanities.

The enhanced ability to imagine and the possibility of projecting that is inherent in the human Umwelt gives rise inevitably to ethics. Ethics requires both the ability to envisage another world more ethical than the present situation. Still, it is important to avoid the assumption that ethics implies agency in the form of will (see Chap. 5, above). The humanist defenders of the humanities seem to suggest that humanities topics cultivate agency in the direction of ethical projects. This, of course, is a gross over-simplification of both structural/agentive interaction and ethics. As has been seen, biosemiotics offers a corrective to the individualist accounts of human agency which subtend the humanist defence of the humanities. As the animal which is distinguished by its ability to recognize that there are such things as *signs* rather than simply responding to signs, Deely et al. (2005) hold that the human is compelled to care for semiosis or, by association, all life on the planet. What circumstances are needed for this to be universally realized has only been hinted at in Chap. 5, above; however, the displacement of human uniqueness to the domain of semiosis from the essence that is beloved of the humanists constitutes a significant step.

Biosemiotics has not been alone in questioning whether ethics represents the pinnacle of a human essence. Posthumanism, zoosemiotics and animal studies have been prominent in dispelling essentialism. Furthermore, they have also been instrumental in asking whether there is some pattern in nature at large whence ethics arises (see, for example, De Waal 1996). If there is such a pattern, humanism's already low stock is further depleted and the defence of humanities needs to look for more rigorous arguments. If the humanities are not the repository of good 'values', if they do not teach people how to live their lives, if they do not directly guarantee the preservation of both democracy and civilization or promote diversity, if they are not inherently 'good', if they do not prevent dehumanization, if they do not exist to shoulder these social roles, then what are the humanities for?

Despite the fact that it is unsupported by a source and noted as apocryphal, Churchwell (2014: 29) nevertheless 'quotes' Richard Dawkins – fanatical promoter of science, arch-mechanist, militant atheist, and the emotional punchbag of all manner of people, from fundamentalist Christians to vitalists – as saying, on exiting an art gallery in Florence, "But what's all this art *for*?" She argues that this question articulates "a widely held view among the instrumentalists and technocrats who decide our society's priorities". Clearly, she does not believe that this is a fair and valid query. There is some sense to this because culture is 'pointless' in the same way as Deacon's (2012a) 'constraints' rest on what is "not there" (see Chap. 7, above). Yet while the apocryphal question is not particularly well put, it is fair to ask it in general terms for the simple reason that the arts and humanities themselves have also always been instrumental. They cannot be defended by humanists as the repository of values one minute and then be pronounced to be value-free the next. Typically, the humanities have been particularly instrumental when they have been denying their instrumentality: at moments of crisis such as the one they are experiencing now, or at moments of triumph when they have served the purposes of colonialism through intellectually subjugating non-Western people. The criticality which exposes such denial is a discourse on instrumentality, as is the meta-criticality which humanists eschew. Furthermore, criticality sees such denial also outside the colonial moment strictly defined; even in the humanities' social tasks, lauded by the humanists, of upholding diversity, multiculturalism, tolerance and gaining local knowledge, there lies instrumentalism and even aggression (see Alibhai-Brown 2000).

A distinct difference characterizes the 'other' humanities, a difference which humanist public relations neglects to mention. Many of the 'other' humanities, without bracketing social issues, have introduced, in varying degrees, questions of cognition and evolution. Diversity, for example, is conceived in the 'other' humanities as a matter of learning the multifarious ways in which the world can be modelled. It is not a matter of discovering the many artifacts accruing to different cultures around the globe and fitting them into a Western definition of universal values. Rather, in posthumanism and animal studies, to take two related instances, diversity entails considering how animality traverses human and non-human worlds and where the human gradually gives way to the machine. Such perspectives would seem to offer much more mileage regarding the question of what the humanities are for than either the affronted response of humanists that the question is indecorous, or the unsustainable assertions that the humanities guarantee the growth of a fictional human essence and the establishment of a utopian 'good' society.

Biosemiotics' perspective on the arts and humanities is, in part, best exemplified by the conclusions from Sebeok's essay, 'Prefigurements of art' (1979b). Published in *Semiotica* as part of a special issue on semiotics of culture edited by Irene Portis-Winner and Jean Umiker-Sebeok, the issue itself derived from the annual meeting of the American Anthropological Association in 1977 and the essay is 70 pages long (including illustrations). It mainly consists of a literature review of extant work on "aesthetic behavior" among non-human animals. It does not take much imagination to see how the essay might sit in the larger book on communication in animals and humans that Sebeok hinted in a number of places that he was writing (for example, in 'Japanese monkey performances' 1986a: 115). If this projected volume could be reconstructed from Sebeok's archived unpublished papers, it is possible that it would assume a status akin to the *Grundrisse* or Freud's 'Project for a scientific psychology'. More so than them, the thesis of 'Prefigurements' is adumbrated and telegraphic, as well as subtle; for that reason, and because of what its importance in biosemiotics and beyond, it is amplified here.

The general purpose of 'Prefigurements of art' is to ask

> whether the optimal design of certain animal communication systems can allow, given certain contextual conditions, for a superimposed aesthetic function. In other words, how reasonable is it to search for prefigurements of aesthetically charged averbal sign configurations in man's animal ancestry? (1979b: 5)

For Sebeok, the idea that the verbal codes of humans simply replaced the averbal systems of animals is untenable and he is sceptical of the idea that the phylogenesis of language can be sought in averbal – Sebeok here uses the term in preference to nonverbal – communication systems (1979b: 8). Yet, this does not rule out the tracing of prefigurements of human averbal aesthetic behaviour in the activities of some animals. This is because human averbal codes have their provenance in the minor hemisphere, "a very superior animal brain", whereas the verbal arts originate in the dominant hemisphere (1979b: 7).

One of the most well-known examples of aesthetic behaviour in animals is the activities of the satinbirds (*Ptylonorhynchus violaceus*) who appear to paint the inside of their bower for purely decorative reasons. It seems that the jury is still out on this issue, with researchers still pursuing the possibility that the painting serves the survival purpose of attracting a mate (Katsuno et al. 2013). Yet, even strict evolutionists in the past – Sebeok cites Thomas Huxley and Dobzhansky – have been willing to embrace the idea that, in such activity, there is "definitely the beginning of aesthetics" (1979b: 6). As Sebeok notes,

> All researches in this field are stamped by a tension between a deeply felt conviction on the part of many distinguished and sensitive biologists that artistic activity indeed exists in the animal world and the inability to face its presumed lack of importance, even uselessness (1979: 30).

By "uselessness", here, Sebeok is referring principally to the lack of a strict survival motive underlying the activity, some means by which the animal's semiosis is not necessarily geared to the preservation of itself and the passing on of its genes. However, it is important to re-visit this idea of "uselessness" and extend it in the process of providing a gloss Sebeok's findings.

For the practical purpose of conveying the knowledge garnered from his literature review, Sebeok identifies four general areas of aesthetic semiosis that have been observed among animals. The first involves kinaesthetic signs, semiosis in movement, particularly as it is envisaged as prefiguring human dance. The second is the realm of musical signs, auditory semiosis that goes beyond communicative calls, even encompassing rudimentary melody, harmony and sequenced repetition. The third comprises pictorial signs, framed visual embellishment which takes place seemingly for its own sake. The final area of semiosis involves architectural signs, semiosis invested in building, beyond the practical requirements of shelter, warmth and protection.

In considering kinaesthetic signs, Sebeok draws attention to now famous examples such as those of the crane, the chimpanzee and birds in the species of the family *Pipridae* (including *el toledo*), the latter of which both sing *and* dance. Dance in animals, he concludes, is homologous with human dance, "much as laughter and smiling fit into the phyletic scale" (1979b: 17). That is to say, dance is passed on through the genome rather than from local and traditional practices. Sebeok is at pains to stress that this does not entail that dance is innate: "information may be communicated to a succeeding generation in several different ways, and therefore, since form depends on the function, convergence can hardly be excluded" (1979b: 18). With this last comment, along with a few others in the article, Sebeok presents an explicitly biosemiotic perspective in stark contrast to a reductive neo-Darwinian one.

The musical signs that seem to amount to aesthetic behaviour in non-human animals are various. As Sebeok notes, an early proposal regarding ornithomusicology stems from a contention of Montaigne that humans first heard birdsong and then went on to imitate it (1979b: 18). Yet, this cannot stand as a pronouncement on the origins of music because of the body of work on musical signs among animals such cicadas, humpback whales and singing gibbons that Sebeok urges us to consider. Despite the complexity of animal musical signs, Sebeok is circumspect in noting that it is premature to take for granted the aesthetic function in such creatures. Extant knowledge about animals' pictorial signs, on the other hand, enable Sebeok to draw more general conclusions. In addition to birds' decoration of nests, he also discusses the finger paintings, reported by Desmond Morris, of the young chimpanzees, Congo, Alpha and Betsy (1979b: 32–4). Not only did the chimps seem to take pleasure in creating their paintings, they also spent time observing their paintings when completed. A later Huxley, Julian, referred to the chimp paintings as evidence of "aesthetic potentialities" among primates.

It is possibly the reports of architectural signs that provide particularly spectacular evidence of "aesthetic potentialities" for Sebeok. He writes (1979b: 43),

> In looking at the endlessly manifold abodes constructed by animals – that serve perhaps to trap prey, to protect or comfort the architect or its kind, especially the young, or to attract the attention of a potential mate – we must look for the artistic value that may be involved, although subordinated to the principal interest of the "survival machine", as Dawkins (1978: 21, 25) calls the temporary receptacles housing the colony of genes inhabiting every plant and animal. If there is such a subsidiary purpose, falling passively under the sway of 'mere' biological advantage, or supplementing it, an effort must be made to ferret out this aesthetic component. Such a quest is far from trivial, for, in the end, it is tantamount to asking: what is art?

The architectural activities of animals, Sebeok argues, are to be understood as manifestations of tool use. Animals have certainly been observed to use tools for specific purposes in a range of different ways. Yet, Sebeok draws attention to those studies which reveal tool use with no apparent specific purpose, citing Frisch to the effect that it cannot be *known* what goes through the animal's mind during such tool use but, nevertheless, again following Frisch, there appears to be evidence of "aesthetic feelings" among such creatures (1979b: 48). A case in point, notes Sebeok, is the beaver, whose skills seem to be innate but also comprise remarkable adaptation to environmental circumstances.

Culture, including aesthetic behaviour, is not just made up of verbal arts. This is an uncontroversial, even trivial, statement because it is so obvious despite the constraint that has been put on nonverbality historically (see Chap. 7, above). What is less obvious and what Sebeok confronts in 'Prefigurements' is whether the averbal arts of humans have an ancestor in the activities of other animals with which humans share this averbal capacity. Yet, more telling still, reflecting the fuzzy 'back and forth' picture of Lamarckian adaptation and Darwinian selection, as well as brain structure, is the advantage that verbality secures for humans. Paradoxically – because non-human animals, of course, possess no fledged faculty resembling language – the survival virtues of verbality present a clue to the purposiveness of 'useless' aesthetic animal behaviour. Language conferred on humans advantages for survival and

> the advantage lay in the extraordinary suppleness of the verbal code – unprecedented in evolution, save for the genetic code to decompose and reconstitute in the human *Innenwelt* the incoming sign vehicles filtered out from the relevant portion of man's environments or *Umwelt*, by our total sensory apparatus. This suppleness is a consequence of the dual organization of the verbal code, which makes it feasible for the human mind to model the world and then, in the fashion of a tinkertoy, to 'play around' with this model: to take it apart, then reassemble it in may different novel arrangements (Sebeok 1986a: 91).

This plasticity is now well known and is a mainstay of much discussion in contemporary cognitive theory. For some time already, Sebeok suggested that grammar allows humans: to posit several putative pasts, to fabricate many kinds of possible future worlds, to imagine death, to create both poetry and science and to project into the future in a fashion that is of a piece with the potential production of an infinite number of sentences from a limited syntax. Above all, perhaps, it allows humans to classify their astonishingly variegated *Umwelt*. Sebeok's conclusion is that animals 'indulge' in aesthetic behaviour because it is a particular form of classifying that is operative within their acts of modelling. As such, aesthetic behaviour *is* an activity with a purpose, even while it seems to have nothing to do with the +, – and 0 of non-human animal existence: the positive seeking out of nourishment/comfort, the avoidance of harm/predation, the general awareness of elements of the environment that can be safely ignored. "The capacity for effective classification", Sebeok writes (41)

> is important for survival, perhaps on a par with eating and sex. If so, techniques of classification were bound to evolve so as to be a source of pleasure to the animal and thus to shape the non-random differential reproduction of its genes (natural selection) ... In other words, although art is always unpredictable, 'it appears to us to have been directed by some organizing center of large codimension, far from the normal structures of ordinary thought, but still in resonance with the main emotional or genetic structures underlying our conscious thought' (Thom 1975: 316).

Aesthetic behaviour, as formulated here, heightens cognitive differentiation. It is a form of modelling with its own specific procedures, practices and rewards. The product of aesthetic behaviour is to simultaneous embellish and furnish animals' niches while also augmenting their basic modelling capacities.

Understanding animals' aesthetic behaviour creates the grounds for understanding the definition of human being. Sebeok juxtaposes the traditional pursuits of philosophical aesthetics with the task of defining life, again problematising pat distinctions between what is learned and what is biologically determined:

> The challenge, of course, is to explicitly define what those relations – of balance and order that delight – are in the characteristic idiom of each art, as well as in the all-embracing architectonics of the living megacosm. The concept of delight thus undergoes a radical transmutation: it is elevated into a function that biologists can recognize, objectify, cope with in familiar terms. The 'artistic animal' is not defined by a heightened sensitivity to movement, sound, color, shape, but by its innate and/or learned capacity to elicit a stable dynamic structure from the fluid environment, whether inorganic, organic, or a subtle blend of both. The sign systems thus created, which serve an underlying semantic function, take in time an aesthetic turn (1979b 58–9).

Sebeok's unravelling of the skeins in which classification and aesthetic behaviour are knit is, it is worth reiterating, subtended by the distinction in modelling between humans and non-human animals, with the latter utilising a limited store of averbal modes and the former having access to an extensive repertoire of both verbal and averbal modes, frequently producing hybrid modes of considerable complexity and nuance. He thus, also, anticipates the arguments about humans as the 'semiotic animal' (Deely 2010).

Humans' knowledge of signs, derived from possession of 'language' or grammar and powers of displacement (see Chap. 2, above), enables projection into the future in addition to humans' highly enhanced ability to classify in the present. It is the anticipation and creation of possible worlds, including those that are fictional. Among the few to have taken up its themes, Johansen (2011) usefully provides an initial list of advances wrought by the human propensity for fictionality:

1. Creating virtual/fictional worlds allows people, in their imagination, to experience what would otherwise be impossible, dangerous, despicable, etc. Hence, such activities may be gratifying by adding a set of wish-worlds to the world we inhabit just because our bodies reside within it.
2. However, creating virtual/fictional worlds does not only satisfy our desires and thrills; it also allows experimentation and model creation, mentally and/or externally, and such alternative versions of parts of our lives, of our interaction, and our action on nature, may not only be consummated individually, they may be communicated.
3. By being communicated, such models may create a common ground that can be shared among the members of a group/society, and hence they may facilitate social integration and communal action.
4. Furthermore, representations of virtual, including counterfactual, scenarios may question existing states of affairs and social relationships, norms, and values.
5. Visions embodied in works of art and literature may attempt to provide answers to questions that are existentially relevant to society and its members.
6. Hence, works of art and literature may either help to strengthen norms, or they may suggest alternative ones.

In 'Prefigurements' (60), Sebeok observes that part of the projection of which fiction is a part derives from mapping and classification capacities of humans:

> It seems clear that the fundamental role of the central nervous system is precisely to provide the creature with a local map simulating its position in the environment, to enable it to sort out, among other vital intelligence, the images of biologically and/or socially important organisms, viz., to distinguish prey from predator. This is surely best accomplished by an arrangement of such images into a distinctive feature matrix, or in terms of 'likeness tempered with difference'.

Effectively, classification enables humans to know their immediate environment and its co-habitants, to enables them to draw maps of extended areas beyond their immediate environment and, ultimately, to function with reference to a cosmology. It also enables the devices of memory, from the complicated spatial mnemonic edifices invented by the ancient Greeks, through the memory palaces of the medieval Jesuits, up to all the storage facilities of post-literate culture. Twenty-three years before Hauser et al. (2002) reached a similar conclusion about the possible basis of language's recursive property in animals' need for navigation, Sebeok found that human mapping exemplifies the constant trafficking between verbal and nonverbal sign systems.

However, it is not the movement between putative different modelling centres in the brain that is the paradox arising from Sebeok's 'Prefigurements'. The ultimate paradox is that animals' aesthetic behaviour is implicated in enhancing survival by not enhancing survival. Much, if not all, of the use-value of aesthetic behaviour consists in not appearing to possess use-value. It is a Deaconian constraint that produces regularity but does so with reference, in survival terms, to something that is not there. Sebeok's isolation of this paradox in the semiosis of animals, including humans, poses a major problem. It is interesting to know that non-human animals are potentially securing their allotted existence whilst furnishing their *Umwelten*; but what are humans doing?

What 'Prefigurements' demonstrates, ineluctably, is that 'Knowledge is for something'. Knowledge was always for something. Knowledge will always be for something. It has a 'scaffolding' dimension for humans (see Cobley and Stjernfelt 2015 and Chap. 4, above); it has other, more direct, functions, too, whether that knowledge is 'scientific', based on putative 'third-person experience', or whether it is knowledge of a different stamp, rooted in 'first-person experience'. In light of Sebeok's essay, the kind of knowledge offered by the humanities could never exist for its own sake, although its layering over might give that impression. As such, aesthetic behaviour *is* survival and it is so because of its contribution to the *Umwelt* of the animal engaged in such behaviour. 'Prefigurements' re-poses the big question for contemporary societies: "What do we pursue in order to maintain an activity which ensures our survival but is not often used instrumentally as such?" As an answer to the question put in this way, research in arts and humanities is to be conceived as a survey of what is being (or has been) explored in the human *Umwelt*, how that has taken place, how the human *Umwelt* is furnished and embellished, and also (where it is possible to discern) what contributions the arts and humanities are making to the survival of the species. *This* is what the humanities are for. They have a cognitive bearing that does not occlude their social bearing. The act of 'aesthetic classification' discussed by Sebeok is the stock-in-trade of the arts. The work of 'aesthetic classification' is central, also, to the humanities in its close relation to the arts.

To preserve the humanities there is a need to be assured of what the humanities are for – that is, not to enable the saccharine sweet appreciation of a human essence but to enhance survival chances in an *Umwelt* that is threatened by despoliation. The humanities' task is to present ways to understand the limits of human agency and its continuity with the agency of other organisms on the planet. It is also to search for ways to grasp the relationship of responsibility entailed in the ethics of this continuity. At present there is a need for the humanities to preserve experience through principally cognitive rather than social means. Eco (2014) warns that humans' reliance on internet-connected devices is causing memory loss; he advises his grandson to start memorizing things for himself from the arts so that he will be fulfilled, as if he had "lived a thousand lives" rather than lived a life "monotonous and devoid of great emotion". Taking away opportunities to interact, participate and analyse the arts represents a threat that should not be underestimated in the age of online memory. It threatens to destroy human experience – first-person experience and knowing – much more rapidly than cyborgs ever could. Accompanying that would be the loss of a sense of the constituents of what it is to be human. As a product of the only animal that can recognize that there are such things as signs, then, it is the task of the humanities not to be party to a limiting of experience and not to allow overspecialization to become an obstacle to presiding, in a Hippocratic manner which does no harm, over the diversity of semiosis.

Conclusion

The main conclusion of this volume is rather conservative in the sense that it merely points to something that is already visible in plain sight. That is, the fact that there is continuity across nature and that humans evolved from other forms of life. Nobody needed this book, or biosemiotics in general, in order to find out that. A walk outside one's office or house or school reveals humans' embeddedness in nature, with plants, animals and insects being encountered in large numbers even while one fails to notice pets inside the house, the microbes in the office or the frenetic endosemiosis within the human body. What is needed, perhaps, is a reminder that all the paraphernalia of culture – the contents of the bookshelves, the online content available through televisions and computers, keepsakes, interior design, utensils, toys and all the signs they comprise – are not only located within the same broad space inhabited by the signs of microbes within and without the human body, but that the palpable signs of culture are also continuous with these identifiably organic signs. Some parts of biosemiotics reveal that to study life is to study signs. To see the world in a grain of sand … The aim of this book has been to emphasize that to study signs is to study life. If none of the other implications of biosemiotics for understanding culture are made clear, hopefully this one will be.

Even the most focused investigation of a cultural phenomenon, blinded to any natural context, is engaged in an analysis of nature. Revealing that and elucidating the context, is another matter. As has been seen, semiotics represents the most meaningful first step towards carrying out this task, although semiotics has not always been able to escape institutional determinations that have impeded the task. While semiotics has carried out work of immeasurable importance in opening out all of culture for analysis, the infusion of biosemiotics prompts semiotics to cast its net so as to analyse sign systems in the whole of nature. The implication is that understanding culture is not just about interrogating how humans operate amid signs, but how they operate amid nature (comprising signs), what distinguishes their cognition and their being as endosemiotic phenomena among other organisms and in the cosmos. As has been seen, what does distinguish humans is the inhabiting of a particular *Umwelt*, a particular kind of modelling comprising nonverbal and verbal

P. Cobley, *Cultural Implications of Biosemiotics*, Biosemiotics 15,
DOI 10.1007/978-94-024-0858-4

modes upon which rest the unique phenomenon of language. The powers of cognitive differentiation that this *Umwelt* affords give rise to culture and make culture what it is. While culture rests on language, the implication is not that culture is to be understood through language alone. Such glottocentrism is a 'forgetting' of the dual verbal/nonverbal heritage of human modelling. The implication for culture is that humans are special because of the way language arises and what it enables humans to do; but humans and their culture cannot be divorced, or the object of a break in continuity, from nature; for even in language humans still carry the heritage of their consanguine natural forms. The ontogenetic and phylogenetic 'forgetting' of humans' nonverbality is not to be underestimated. At the same time, the contribution of nonverbality to processes of 'knowing', as well as considering aspects of culture as kinds of 'knowing' rather than attributing that faculty to Φ-sciences alone, is another implication of biosemiotics (and cybersemiotics) for culture and cultural analysis.

At the level of consideration of humans' position within sociality and cultural exchange, biosemiotics has further implications for culture. Those positions in the analysis of culture and in the popular imagination which posit individualism, sometimes along with the assertion that humans are exceptional, are especially undermined by biosemiotics. Both humanism and individualism, separately and together, are rendered incompatible with biosemiotics' repeated finding of continuity of signs and continuity across nature. Evidence of humans' collective bearing – from endosemiosis, through early evolution of hominids, in elementary social groupings or swarms, through to complex societies – is always on the table for biosemiotics. So, while biosemiotics identifies agency and degrees of freedom across the realms of nature, seemingly making it compatible with some perspectives on culture which eschew dun scientism and exalt agency in culture, that compatibility is very much qualified – to the point of being illusory. Put another way, 'agency' does not correspond to individualism; agency takes place within constraints and within collectivity.

This is not to forget that biosemiotics also has implications for those more radical perspectives which are sceptical of agency and see humans as subjects, traversed by culturally constructed power relations. Cultural construction, here, often implies 'language' or 'construction in discourse'. However, as has been seen, the conception of 'sign' that informs biosemiotics far surpasses the explanatory power of 'discourse'. It not only covers nonverbal signification and semiosis beyond the human, but also carries with it an understanding of how human signs fluctuate between mind-dependency (construction) and mind-independence. Hence, for the natural subject, living wholly (non-human animals) or partly (humans) in a world of objects, the other is both everything and oneself. The implications of this, for the temptations that might exist in current understandings to maintain the binary of self/other – as well as individual/collectivity, agent/subject, verbal/nonverbal, non-human/human, matter/mind, living nature/culture – are clear: not only are the binaries false oppositions but the human *Umwelt*, characterized by its constant drive to expand its range, should not allow such binaries to hinder its enrichment. As Peirce (1.135) would say, "Do not block the way of inquiry". That culture and the study of culture have the foremost role to play in this should go without saying.

Accompanying this implication is the observation that the natural ways of culture, once more, are often overlooked. Ethics, seemingly the product of good will and the programmes that result from such will, has been shown to be a by-product with a natural grounding. Necessarily, this cultural implication is more speculative in that there is much of practical ethics that does rely on a willed programme at present. The biosemiotic conception of ethics, exemplified by parenthood, is no doubt everywhere in nature, begetting further signs of ethics. The implication for culture is that there is a need to renew the search for ethics in these terms.

Sometimes programmes for ethics are called 'ethical' codes because they promote putatively inviolable instructions. In ethics, as elsewhere in culture, codes cannot be relied upon to work in this mechanical way. Nor should it be assumed that there are codes in nature that act as a model for this kind of mechanism in culture, however much this volume has been dedicated to elucidating biosemiotic continuity across nature. As Hoffmeyer and Emmeche (2007: 51; emphasis in original), warn in relation to the assumption that culture is a by-produce of germ cells, "*The semiotics of nature should not be confused with the semiotics of human culture*". The key point in emphasizing continuity is the *type* of activity involved rather than the *token*. As has been stated, not all coding is like the human invention called cryptography. If anything in respect of codes is continuous across nature (and therefore in culture) it is their fallibility. Not all understandings of culture posit the strong efficacy of codes, even while the idea has a grip on the popular imagination; at the same time, not all understandings of culture insist that culture is without codes or that it is completely individual, another idea with a strong grip on the popular imagination. The implication of the discussion of codes in this book is, predictably, that understandings of culture might be enhanced if the processes of invariance – as opposed to mechanical coding – that appear in nature were seen in continuity with like processes in the compartment of nature called culture. Biosemiotics has an extremely useful and illuminating approach to invariance that it has inherited from Peirce: the habit. However, a more refined understanding of the process, still, is offered by Deacon in the biosemiotic idea of 'constraint'. This also promises to account for what is lost as well as what remains in natural development.

The issue of what is lost in culture is no doubt a perennial one. Culture is always 'in crisis'. However, there are cogent reasons to assume that humans are at a critical point in the early decades of the twenty first century. As has been seen, social formations are preparing to organize themselves for a future where technology will transform traditional forms of culture. This is a charitable stating of the case. Another way of describing the matter is to say that crass instrumentalists are dismantling the arts and the humanities that study them at the very moment when technologies are emerging that those instrumentalists can reference in their myopic cost-cutting exercises. To state the matter in this way does not amount to Luddism. The implication from biosemiotics, here, concerns the shearing of humans' instruments of cognitive differentiation and the out-sourcing of memory and experience. The implication, put bluntly, is that aesthetic behaviour *is survival* – it locates humans in their world and enables humans to conceptualise the furnishing of that world. It has indispensable use-value and, in fact, underpins the very science and technology that economic instrumentality seeks to exploit. Foresight, to recognize how seemingly non-purposive signs enhance the *Umwelt*, is paramount, as well as analytic acumen in understanding the relation of aesthetic signs to human existence in the past, the present and the future. In sum, addressing the big question of aesthetic behaviour requires experienced, interdisciplinary technicians to be centrally involved. There can be no equivocation on this. Naturally, when the long-term benefits of an activity are obscure, there is difficulty in making them the core of a consensus. Yet, such difficulties are not insurmountable, particularly as the topic under discussion is one that cuts to the very nature of humans' existence and their most 'instinctive' pursuits. Sebeok (1979b: 42) notes, with a nod to Peirce,

> The propensity to classify seems to have acquired, through evolution, diminishing survival value, but then so did sex: humans can enjoy either, but most tokens, though pleasurable per se, are not biologically relevant. Only the type of activity has a clearcut biological function.

Although restaurants have garnered business for centuries, and prostitution has flourished for longer still, nobody has suggested that eating and sex should only ever be undertaken for the purposes of generating profit. Possibly the ultimate cultural implication of biosemiotics is that the absurd proposal for only ever engaging in aesthetic behaviour for brute economic purposes should be banished from the realms of common understanding.

References

Adler M. The difference of man and the difference it makes. New York: Holt Rhinehart; 1967.
Agamben G. In: Attell K, editor. State of exception. Chicago: Chicago University Press; 2005.
Agamben G. Homo sacer. Trans. Heller-Roazen D. Stanford: Stanford University Press; 1998.
Alibhai-Brown Y. After multiculturalism. London: Foreign Policy Centre; 2000.
Alibhai-Brown Y. After multiculturalism. London: Foreign Policy Centre; 2002.
Althusser L. For Marx. Trans. Brewster B. London: New Left Books; 1969.
American Council of Learned Societies. Text of 'Speaking for the Humanities'. Chron High Educ. 1988;35(A1):A11–2.
Andrews KR. Liberal education for competence and responsibility. In: Donaldson TJ, Freeman ER, editors. Business as a humanity. New York: Oxford University Press; 1994. p. 152–66.
Ascher M, Ascher R. Mathematics of the Incas: code of the Quipu. New York: Dover; 1997.
Atkins K. Introduction. In: Atkins, editor. Self and subjectivity. Oxford: Blackwell; 2005.
Augustyn P. Uexküll, Peirce, and other affinities between biosemiotics and biolinguistics. Biosemiotics. 2009;2(1):1–17.
Badiou A. Ethics: an essay on the understanding of evil. Trans. Hallward P. London: Verso; 2001.
Baldwin J. Beyond bodies, culture and language: an introduction to Alain Badiou. Subject Matters. 2004;1(2):1–20.
Barbieri M. Has biosemiotics come of age? Semiotica. 2002;139(1/4):283–95.
Barbieri M. The organic codes: an introduction to semantic biology. Cambridge: Cambridge University Press; 2003.
Barbieri M, editor. Introduction to biosemiotics: the new biological synthesis. Dordrecht: Springer; 2007a.
Barbieri M, editor. The codes of life: the rules of macroevolution. Dordrecht: Springer; 2007b.
Barbieri M. A short history of biosemiotics. Biosemiotics. 2009;2(2):221–45.
Barbieri M. On the origin of language: a bridge between biolinguistics and biosemiotics. Biosemiotics. 2010;3(2):201–23.
Barry B. Culture and equality. Oxford: Polity; 2000.
Barthes R. Mythologies. Trans. A. Lavers, London: Paladin; 1973.
Barthes R. From work to text. In: Stephen Heath, ed. and trans. Image – music – text, London: Fontana; 1977a. p. 155–64.
Barthes R. Change the object itself. In: Stephen Heath, ed. and trans. Image – music – text, London: Fontana; 1977b.
Barthes R. Elements of semiology. Trans. Lavers A. and Smith C. London: Cape; 1967, [1964].
Bartlett F. Remembering. Cambridge: Cambridge University Press; 1932.

P. Cobley, *Cultural Implications of Biosemiotics*, Biosemiotics 15,
DOI 10.1007/978-94-024-0858-4

Bate J. Introduction. In: Bate J, editor. The public value of the humanities. London: Bloomsbury; 2011. p. 1–16.
Beard M. Live classics or 'What's the use of Aeschylus in Darfur'? In: Bate J, editor. The public value of the humanities. London: Bloomsbury; 2011. p. 17–29.
Beever J. Meaning matters: the biosemiotic basis of bioethics. Biosemiotics. 2012;5:181–91.
Beever J. and Tønnessen M. Justifying moral standing by biosemiotic particularism. Zeitschrift fur Semiotik. 2016; (forthcoming).
Belsey C. Culture and the real. London: Routledge; 2005.
Benhabib S. Situating the self: gender, community and postmodernism in contemporary ethics. Cambridge: Polity; 1992.
Bennis WG, O'Toole J. How business schools lost their way. Harv Bus Rev. 2005;83:96–104.
Bentham J. In: Božovič M, editor. The panopticon writings. London: Verso; 1995.
Biggs NL. An introduction to information communication and cryptography. Dordrecht: Springer; 2008.
Birdwhistell RL. Kinesics and context. Philadelphia: University of Pennsylvania Press; 1970.
Blasco M. Cultural pragmatists? Student perspectives on learning culture at a business school. Acad Manag Learn Educ. 2009;8(2):174–87.
Bloom C. Violent London: 2000 years of riots, rebels and revolts. London: Sidgwick and Jackson; 2003.
Bourdieu P. Distinction. Trans. Nice R. London: Routledge; 1984.
Bragg M. The adventure of English: the biography of a language. London: Sceptre; 2003.
Brent J. Charles Sanders Peirce: a life. Bloomington: Indiana University Press; 1993.
Brentari C. Jakob von Uexküll: the disovery of the umwelt between biosemiotics and theoretical biology. Dordrecht: Springer; 2015.
Brier S. Cybersemiotics: why information is not enough! Toronto: University of Toronto Press; 2008a.
Brier S. The paradigm of peircean biosemiotics. Signs. 2008b;2:20–81.
Bruner JS. Going beyond the information given. New York: Norton; 1957.
Bruner JS. The process of education. Cambridge, MA: Harvard University Press; 1960.
Bruner JS. Toward a theory of instruction. Cambridge, MA: Belknap; 1966.
Bryson B. Mother tongue: the English language. Harmondsworth: Penguin; 1990.
Bryson JS. How Jihadist education breeds violence Public Discourse. (2010, 29 Mar). http://www.thepublicdiscourse.com/2010/03/1221/. Last accessed 17 Nov 2014.
Bühler K. The theory of language: the representational function of language. Amsterdam: John Benjamins; 1990.
Burgess A. A mouthful of air: language and languages, especially English. London: Vintage; 1993.
Cannizzaro S, Cobley P. Biosemiotics, politics and Sebeok's move from linguistics to semiotics. In: Velmezova E, Cowley SJ, Kull K, editors. Biosemiotic perspectives in linguistics. Dordrecht: Springer; 2015.
Cascardi AJ. The subject of modernity. Cambridge: Cambridge University Press; 1992.
Cheney DL, Seyfarth RM. How monkeys see the world: inside the mind of another species. Chicago: University of Chicago Press; 1990.
Chomsky N. A review of B. F. Skinner's verbal behavior. Language. 1959;35(1):26–58.
Chomsky N. Aspects of the theory of syntax. Cambridge, MA: MIT Press; 1965.
Churchwell S. Humanities: why the study of human culture still matters. Times High Educ. 13 Nov 2014.
Cobley P. The American thriller: generic innovation and social change in the 1970s. London: Palgrave; 2000.
Cobley P. How to be evil. Subject Matters. 2004;1(2):47–66.
Cobley P. Culture: definition and concepts. In: Donsbach W, editor. International encyclopedia of communication. New York: WileyBlackwell; 2008. p. 1127–32.
Cobley P, editor. Realism for the 21st century: a John Deely reader. Scranton/London: University of Scranton Press; 2009.

Cobley P. Sebeok's panopticon. In: Cobley et al., editors. "Semiotics Continues to Astonish": how Thomas A. Sebeok shaped the future of the doctrine of signs. Berlin: de Gruyter; 2011. p. 85–114.

Cobley P. Metaphysics of wickedness. In: Thellefsen T, Sørensen B, editors. Peirce in his own words. Berlin: de Gruyter; 2014.

Cobley P. The deaths of semiology and mythoclasm: Barthes and media studies. Signs Media. 2015;10:1–25.

Cobley P, Randviir A. What is sociosemiotics? Sociosemiotica special issue eds. Cobley and Randviir Semiotica. 2009; 173 (1–2):1–39.

Cobley P, Stjernfelt F. Scaffolding development and the human condition. Biosemiotics. 2015;8(2):291–304.

Cobley P, et al. editors. Semiotics continues to astonish: Thomas A. Sebeok and the doctrine of signs, Berlin: de Gruyter Mouton, 2011.

Cobley P, Deely J, Kull K, Petrilli S. Introduction: Thomas A. Sebeok: biography and 20th century role. In: Paul Cobley et al., editors. "Semiotics Continues to Astonish": Thomas A. Sebeok and the doctrine of signs. Berlin: De Gruyter Mouton; 2011.

Colapietro VM. Peirce's approach to the self: a semiotic perspective on human subjectivity. Albany: SUNY Press; 1989.

Colapietro VM. Human Understanding in light of John Deely, Charles Darwin, and Charles Peirce. In: Cobley P editor. Deely in Review special issue of Chinese Semiotic Studies 2016; 12 (3): 1–138.

Colby A et al. *Rethinking undergraduate business education: liberal learning for the profession*. Report of the Carnegie Foundation for the Advancement of Teaching. San Francisco: Jossey-Bass; 2011.

Collini S. What are universities for? Harmondsworth: Penguin; 2012.

Csányi V. The brain's models and communication. In: Sebeok TA, Umiker-Sebeok J, editors. Biosemiotics: the semiotic web 1991. Berlin: Mouton de Gruyter; 1992. p. 27–44.

Darwin C. The origin of species by means of natural selection, or the preservation of favoured races in the struggle for life. London: John Murray; 1859.

Darwin C. The descent of man, and selection in relation to sex. London: John Murray; 1871.

Darwin C. The origin of species by means of natural selection, or the preservation of favoured races in the struggle for life. 6th ed. London: John Murray; 1872.

De George RT. Business as a humanity: a contradiction in terms. In: Donaldson TJ, Freeman ER, editors. Business as a humanity. New York: Oxford University Press; 1994. p. 11–26.

de Saussure F. Cours de linguistique générale. Paris: Payot; 1916.

de Saussure F. Course in general linguistics. Trans. Wade Baskin, New York: McGraw-Hill; 1959, [1916]

de Saussure F. Course in general linguistics. Trans. Harris R. London: Duckworth; 1983, [1916].

De Waal F. Good natured: the origins of right and wrong in humans and other animals. Cambridge, MA: Harvard University Press; 1996.

De Waal F. Are we smart enough to know how smart animals are? Kindle Edition ed. London: Granta; 2016.

De Waal F, editor. The tree of origin: what primate behavior can tell us about human social evolution. Cambridge, MA: Harvard University Press; 2001.

De Waal F, Macedo S, Ober J, editors. Primates and philosophers: how morality evolved. Princeton: Princeton University Press; 2006.

Deacon TW. The symbolic species: the co-evolution of language and the human brain. Harmondsworth: Penguin; 1997.

Deacon TW. The hierarchic logic of emergence: untangling the interdependence of evolution and self-organization. In: Weber BH, Depew DJ, editors. Evolution and learning: the Baldwin effect reconsidered. Cambridge, MA: MIT Press; 2003.

Deacon TW. Incomplete nature: how mind emerged from matter. New York: Norton; 2012a.

Deacon TW. Beyond the symbolic species. In: Schilhab T et al., editors. The symbolic species evolved. Dordrecht: Springer; 2012b. p. 1–38.

Deacon TW. The emergent process of thinking as reflected in language processing (not yet published). 2016.

Deacon TW. Consciousness is a matter of constraint. New Sci. 2011. Issue 2840. https://www.newscientist.com/article/mg21228406-300-consciousness-is-a-matter-ofconstraint/. Last accessed 21 Mar 2016.

Deacon TW, et al. Theses on biosemiotics: prolegomena to a theoretical biology. Biol Theory. 2009;4(2):167–73.

Deely J. Introducing semiotic: its history and doctrine. Bloomington: Indiana University Press; 1981.

Deely J. The human use of signs, or elements of anthroposemiotics. Lanham: Rowman and Littlefield; 1994.

Deely J. Umwelt. Semiotica. 2001;134(1/4):125–35.

Deely J. The quasi-error of the external world. Cybern Hum Know. 2003;10(1):25–46.

Deely J. Dramatic reading in three voices: 'A Sign Is What?'. Am J Semiotics. 2004;20(1–4):1–66.

Deely J. The semiotic animal: a postmodern definition of human being to supersede the modern definition as 'res cogitans'. Sofia: New Bulgarian University; 2005.

Deely J. Purely objective reality. Berlin: de Gruyter Mouton; 2009a.

Deely J. Semiosis and Jakob von Uexküll's concept of umwelt. In: Cobley P, editor. Realism for the 21st century: a John Deely reader. Scranton: Scranton University Press; 2009b. p. 239–58.

Deely J. Semiotic animal. South Bend: St Augustine's Press; 2010.

Deely J. Building a scaffold: semiosis in nature and culture. Biosemiotics. 2015a;8:341–60.

Deely J. Semiosis and 'meaning as use': the indispensability and insufficiency of subjectivity in the action of signs. Sign Syst Stud. 2015b;43(1):7–28.

Deely J. Ethics and the semiosis-semiotics distinction. Zeitschrift fur Semiotik. 2016; (forthcoming)

Deely J, Petrilli S, Ponzio A. The semiotic animal. New York: Legas; 2005.

Delbrück M. Mind from matter? Palo Alto: Blackwell; 1986.

Deutscher G. Through the language glass: how words colour your world. London: Heinemann; 2010.

Donald M. Origins of the modern mind. Boston: Harvard University Press; 1991.

Dunant S, editor. The war of the words: the political correctness debate. London: Virago; 1994.

Eagleton T. Figures of dissent: critical essays on Fish, Spivak, Žižek and others. London: Verso; 2003.

Eco U. A theory of semiotics. Bloomington: Indiana University Press; 1976.

Eco U. Unlimited semiosis and drift. In The limits of interpretation. Bloomington: Indiana University Press; 1990. p. 23–43.

Eco U. Caro nipotino mio. L'espresso. 3 Jan 2014. http://espresso.repubblica.it/visioni/2014/01/03/news/umberto-eco-caro-nipote-studia-a-memoria-1.147715?refresh_ce (last accessed 16 July 2016)

Elkins J. Visual studies: a skeptical introduction. London: Taylor and Francis; 2003.

Elliott A. Concepts of the self. Cambridge: Polity; 2001.

Emmeche C. The garden in the machine: the emerging science of artificial life. Princeton: Princeton University Press; 1994.

Engels F. Ludwig Feuerbach and the end of classical German philosophy. Moscow: Progress; 1946 [1886].

Esposito JL. Synechism: the keystone of Peirce's metaphysics. In Digital encyclopedia of Charles S. Peirce, eds. João Queiro and Ricardo Gudwin (n.d.). http://www.digitalpeirce.fee.unicamp.br/p-synesp.htm.

Evans N, Levinson S. The myth of language universals: language diversity and its importance for cognitive science. Behav Brain Sci. 2009;2(5):429–48.

Fant GC, Halle M, Jakobson R. Preliminaries to speech analysis. Cambridge, MA: MIT Press; 1952.

Fast J. Body language. New York: Simon and Schuster; 1970.

Favareau D. Introduction: an evolutionary history of biosemiotics. In: Favareau D, editor. Essentials of biosemiotics. Dordrecht: Springer; 2010a. p. 1–80.

Favareau D, editor. Essentials of biosemiotics. Dordrecht: Springer; 2010b.
Favareau D, et al. The glossary project: intentionality. Biosemiotics (forthcoming).
Favareau D, Kull K. On biosemiotics and its possible relevance to linguistics. In: Velmezova E, Cowley SJ, Kull K, editors. Biosemiotic perspectives in linguistics. Dordrecht: Springer; 2015.
Feldman DC. The food's no good and they don't give us enough: reflections on Mintzberg's critique of MBA education. Acad Manag Learn Educ. 2005;4(2):217–20.
Fish SE. Will the humanities save us? New York Times. 6 Jan 14. 2008.
Florkin M. Concepts of molecular biosemiotics and molecular evolution. In: Florkin AM, Stotz EH, editors. Comprehensive biochemistry, vol. 29. Amsterdam: Elsevier; 1974.
Foucault M. Discipline and punish: the birth of the prison. Trans. Sheridan A. Harmondsworth: Penguin; 1977.
Freeman ER. Epilogue. In: Donaldson TJ, Freeman ER, editors. Business as a humanity. New York: Oxford University Press; 1994. p. 215–26.
Freud S. Repression. In On metapsychology. The theory of psychoanalysis. (Pelican Freud Library 11.) ed. Angela Richards, trans. James Strachey. Harmondsworth: Penguin; 1984[1915]. p. 139–58.
Gambetta D, Hertog S. Engineers of Jihad sociology working papers, University of Oxford; 2007. p. 1–88.
Gaukroger S. The natural and the human. Science and the shaping of modernity 1739–1841. Oxford: Oxford University Press; 2016.
Ghoshal S. Bad management theories are destroying good management practice. Acad Manag Learn Educ. 2005;4(1):75–91.
Giddens A. Modernity and self-identity: self and society in the late modern age. Cambridge: Polity; 1991.
Gordon RA, Howell JE. Higher education for business. New York: Columbia University Press; 1959.
Gould SJ, Lewontin RC. The spandrels of San Marco and the Panglossian paradigm: a critique of the adaptationist programme. Proc R Soc Lond. 1979;B205(1161):581–98.
Gould SJ, Vrba ES. Exaptation: a missing term in the science of form. Paleobiology. 1982;8:4–15.
Gouldner, AW. The two Marxisms. New York: Oxford University Press; 1980.
Greenberg J, editor. Universals of language. Cambridge, MA: MIT Press; 1963.
Guiraud P. Semiology. London: Routledge and Kegan Paul; 1975.
Habermas J. The structural transformation of the public sphere: an inquiry into a category of bourgeois society. Trans. Burger T. and Lawrence F. Cambridge, MA: MIT Press; 1989.
Hall ET. The hidden dimension. New York: Anchor Books; 1966.
Hall JA, Knapp ML, editors. Nonverbal communication. Berlin: de Gruyter; 2013.
Halliday MAK. Language as social semiotic. In Language as social semiotic. London: Edward Arnold; 1978. p. 108–26.
Han L. Umwelt as a female Taoist principle: re-reading the Tao Te Ching. Paper delivered at the 16th Gatherings in Biosemiotics. Prague: Charles University; 2016.
Harries-Jones P. Upside-down gods. Gregory Bateson's world of difference. New York: Fordham University Press; 2016.
Harris R. The language myth. London: Duckworth; 1981.
Harris R. The language connection: philosophy and linguistics. Bristol: Thoemmes; 1996.
Harris R. Introduction to integrational linguistics. Oxford: Pergamon; 1998.
Harris R. Integrational linguistics and the structuralist legacy. Lang Commun. 1999;19:45–68.
Harris R. Saussure and his interpreters. 2nd ed. Edinburgh: Edinburgh University Press; 2003.
Harris R. Integrationist notes and papers 2003–2005. Crediton: Tree Tongue; 2006.
Hauser MD et al. The faculty of language: what is it, who has it and how did it evolve? Science. 22 Nov 2002;298(5598):1569–79.
Hockett CR. The problem of universals in language. In: Greenberg J, editor. Universals of language. Cambridge, MA: MIT Press; 1963.

Hoffmeyer J. Some semiotic aspects of the psycho-physical relation: the endo-exosemiotic boundary. In: Sebeok TA, Umiker-Sebeok J, editors. The semiotic web 1991: biosemiotics. Berlin: Mouton de Gruyter; 1992. p. 101–22.

Hoffmeyer J. The swarming cyberspace of the body. Cybern Hum Know. 1995;3(1):16–25.

Hoffmeyer J. Signs of meaning in the universe. Trans. Haveland B. J. Bloomington: Indiana University Press; 1996.

Hoffmeyer J. Surfaces inside surfaces. On the origin of agency and life. Cybern Hum Know. 1998;5:33–42.

Hoffmeyer J. Order out of indeterminacy. Semiotica. 1999;127(1–4):321–44.

Hoffmeyer J. S/E = 1: a semiotic understanding of bioengineering. Sign Syst Stud. 2001;29(1):277–90.

Hoffmeyer J. Semiotic scaffolding of living systems. In: Barbieri M, editor. Introduction to biosemiotics. Berlin: Springer; 2007. p. 149–66.

Hoffmeyer J. Biosemiotics. An examination into the signs of life and the life of signs. Scranton: Scranton University Press; 2008a.

Hoffmeyer J, editor. A legacy for living systems: Gregory Bateson as a precursor to biosemiotics. Dordrecht: Springer; 2008b.

Hoffmeyer J. Semiotic freedom: an emerging force. In: Davis P, Gregersen NH, editors. Information and the nature of reality. From physics to metaphysics. Cambridge: Cambridge University Press; 2010a. p. 185–204.

Hoffmeyer J. Semiotics of nature. In: Cobley P, editor. The Routledge companion to semiotics. London: Routledge; 2010b.

Hoffmeyer J. Astonishing life. In: Cobley P et al., editors. Semiotics continues to astonish: Thomas A. Sebeok and the doctrine of signs. Berlin: de Gruyter Mouton; 2011.

Hoffmeyer J, Emmeche C. Biosemiotica II, Spec Issue Semiotica. 1999;127(1–4):133–655.

Hoffmeyer J, Kull K. Baldwin and biosemiotics: what intelligence is for. In: Weber BH, Depew DJ, editors. Evolution and learning: the Baldwin effect reconsidered. Cambridge, MA: MIT Press; 2003.

Hoffmeyer J, Emmeche, C. Code-duality and the semiotics of nature. Revised version. In Biosemiotics: information, codes and signs in living systems. Barbieri M, editor. New York: Nova; 2007.

Howard D. Architectural history in academia and the wider community. In: Bate J, editor. The public value of the humanities. London: Bloomsbury; 2011. p. 76–86.

Huyssen A. After the great divide: modernism, mass culture, postmodernism. Bloomington: Indiana University Press; 1986.

Ibrus I. Culture is becoming more visible and therefore richer. In: Development of Estonian cultural space. Tallinn: Estonian Development Report; 2014/2015. p. 222–5.

Ingram J. Talk, talk, talk. New York: Penguin; 1992.

Jakobson R. Linguistics and poetics: closing statement. In: Sebeok TA, editor. Style in language. Cambridge, MA: MIT Press; 1960.

Jakobson R. Six lectures on sound and meaning. Hassocks: Harvester; 1976.

Jakobson R. Some questions of meaning. In: Waugh LR, Monville-Burston M, editors. On language. Cambridge, MA: Harvard University Press; 1990a.

Jakobson R. Quest for the essence of meaning. In: Waugh LR, Monville-Burston M, editors. On language. Cambridge, MA: Harvard University Press; 1990b.

Jakobson R. Langue and parole: code and message. In: Waugh LR, Monville-Burston M, editors. On language. Cambridge, MA: Harvard University Press; 1990c.

Jameson F. Signatures of the visible. London: Routledge; 1990.

Jay M. Downcast eyes: the denigration of vision in twentieth-century French thought. Berkeley: University of California Press; 1993.

Johansen JD. Semiotics, biology, and the adaptionist theory of literature. In: Cobley P, editor. Semiotics continues to astonish'. Thomas A. Sebeok and the doctrine of signs. Berlin: de Gruyter Mouton; 2011. p. 207–22.

Kagan J. The three cultures: natural sciences, social sciences and the humanities in the 21st century. Cambridge: Cambridge University Press; 2009.

Katsuno Y, Eguchi K, Noske RA. Preference for and spatial arrangement of decorations of different colours by the Great Bowerbird. Ptilonorhynchus Nuchalis Nuchalis Aust Field Ornithol. 2013;30(1):3–13.

Kauffman SA. Investigations. New York: Oxford University Press; 2000.

Kean TH et al. The 9/11 Commission report: final report of the National Commission on Terrorist Attacks upon the United States. New York/London: Norton; 2004.

Kelly P, editor. Multiculturalism reconsidered: culture and equality and its critics. Oxford: Polity; 2002.

Kelly M. Language matters 2. Modern languages. In: Bate J, editor. The public value of the humanities. London: Bloomsbury; 2011. p. 259–71.

Kendon A. Gesture: visible action as utterance. Cambridge: Cambridge University Press; 2004.

Kress GR. Against arbitrariness: the social production of the sign as a foundational issue in critical discourse analysis. Discourse Soc. 1993;4(2):169–91.

Kuhn TS. The structure of scientific revolutions. 2nd ed. Chicago/London: University of Chicago Press; 1970.

Kull K. On semiosis, umwelt and semiosphere. Semiotica. 1998;120(3/4):299–310.

Kull K, editor. Jakob von Uexküll: a paradigm for biology and semiotics special issue. Semiotica134 (1/4); 2001.

Kull K. Uexküll and the post-modern evolutionism. Sign Syst Stud. 2004;32(1/2):99–114.

Kull K. A brief history of biosemiotics. In: Barbieri M, editor. Biosemiotics: information, codes and signs in living systems. New York: Nova; 2007.

Kull K. Biosemiotics: to know, what life knows. Cybern Hum Know. 2009;16(3–4):81–8.

Kull K. Umwelt and modeling. In: Cobley P, editor. The Routledge companion to semiotics. London: Routledge; 2010.

Kull K. Advancements in biosemiotics: where we are now in discovering the basic mechanisms of meaning-making. In: Rattasepp S, Bennett T, editors. Gatherings in biosemiotics. Tartu: Tartu University Press; 2012.

Kull K. Adaptive evolution without natural selection. Biol J Linn Soc. 2014a;112:287–94.

Kull K. Zoosemiotics is the study of animal forms of knowing. Semiotica. 2014b;198:47–60.

Kull K, et al. When culture supports biodiversity: the case of the wooded meadow. In: Roepstorff A et al., editors. Imagining nature: practices of cosmology and identity. Aarhus: Aarhus University Press; 2003.

Kull K, Emmeche C, Favareau D. Biosemiotic questions. Biosemiotics. 2008;1:41–55.

Kull K, et al. Theses on biosemiotics. Biol Theory. 2009;4(2):167–73.

Kull K, Velmezova E. Umberto Eco on biosemiotics. Paper presented at the 16th Gatherings in Biosemiotics. Prague: Charles University; 7 July 2016.

Levin D. Keeping Foucault and Derrida in sight: panopticism and the politics of subversion. In: Sites of vision: the discursive construction of sight in the history of philosophy. Cambridge, MA: MIT Press; 1997.

Lotman J. The sign mechanism of culture. Semiotica. 1974;12(4):301–5.

Lotman YM. The text and the structure of its audience. New Lit Hist. 1982;14(1):81–7.

Lovelock J. Gaia: a new look at life on earth. 2nd ed. Oxford: Oxford University Press; 2000.

Lyotard J-F. The postmodern condition: a report on knowledge. Manchester: Manchester University Press; 1984

Machin D. Introduction. In: Machin, editor. Visual communication. Berlin: de Gruyter; 2014. p. 3–22.

Maher J. Seeing language in sign: the work of William C. Stokoe. Washington: Gallaudet University Press; 1997.

Maran T. Semiotic interpretations of biological mimicry. Semiotica. 2007;167–1(4):223–48.

Maran T. Why was Thomas A. Sebeok not a cognitive ethologist? From 'animal mind' to 'semiotic self'. Biosemiotics. 2010;3(3):315–29.

Markoš A. Readers of the book of life. Oxford: Oxford University Press; 2002.

Markoš A, Grygar F, Hajnal L, Kleisner K, Kratochvil Z, Neubauer Z. Life as its own designer: Darwin's origin and western thought. Dordrecht: Springer; 2009.

Marler P. The logical analysis of animal communication. J Theor Biol. 1961;1:295–317.

Marrone G. The invention of the text. Milan: Mimesis; 2014.

Matthews PH. Grammatical theory in the United States: from Bloomfield to Chomsky. Cambridge: Cambridge University Press; 1993.

McAllister QO. Business executives and the humanities. Raleigh: University of North Carolina Press; 1951.

McCrum R, et al. The story of English. rev. ed. Harmondsworth: Penguin; 2002.

Mcdonald R. The value of art and the art of evaluation. In: Bate J, editor. The public value of the humanities. London: Bloomsbury; 2011. p. 283–94.

McEliece RJ. Theory of coding and information. 2nd ed. Cambridge: Cambridge University Press; 2002.

Merrell F. How signs proliferate. In: Signs for everybody: Chaos Quandaries and communication. Toronto: Legas; 2000.

Mintzberg H. Managers not MBAs: a hard look at the soft practice of managing and management development. San Francisco: Berrett-Koehler; 2004.

Mirzoeff N. An introduction to visual culture. London: Routledge; 1999.

Mitchell WJT. Picture theory: essays on verbal and visual representation. Chicago: University of Chicago Press; 1994.

Mongré P. Sant' Ilario. Gedanken aus der Lanschaft Zarathustras. Leipzig: C. G. Naumann; 1897.

Morsing M, Rovira AS. Prologue: business schools as usual? In: Morsing, Rovira, editors. Business schools and their contribution to society. Los Angeles: Sage; 2011. p. xviii–xxi.

Neimark J, Ake S. Consciousness blows my mind: a Stu Kauffman interview. 2002. Metanexus http://www.metanexus.net/Magazine/ArticleDetail/tabid/68/id/5605/Default.aspx. Accessed 11 Mar 2010.

Nelson JK, Poms LW, Wolf PP. Developing efficacy beliefs for ethics and diversity management. Acad Manag Learn Educ. 2012,11(1):49–68.

Nöth W. Semiotics for biologists. In: Barbieri M, editor. Biosemiotics: information, codes and signs in living systems. New York: Nova; 2007.

Nussbaum MC. Not for profit: why democracy needs the humanities. Princeton University Press: Princeton; 2010.

O'Brien R. Foreword. In: Nussbaum MC, editor. Not for profit: why democracy needs the humanities. Princeton: Princeton University Press; 2010. p. ix–xii.

O'Gorman F. Making meaning: literary research in the twenty-first century. In: Bate J, editor. The public value of the humanities. London: Bloomsbury; 2011. p. 272–82.

Pablé A, Hutton C. Signs, meaning and experience: integrational approaches to linguistics and semiotics. Berlin: de Gruyter; 2015.

Parker Pearson M. The value of archaeological research. In: Bate J, editor. The public value of the humanities. London: Bloomsbury; 2011. p. 30–43.

Peirce CS. Guessing. Hound Horn. 1929;2(3):267–82.

Peirce CS. The collected papers of Charles Sanders Peirce. vols. I–VI [Charles Hartshorne and Paul Weiss eds.], vols. VII–VIII [Arthur W. Burks ed.] Cambridge, MA: Harvard University Press; 1931–1958.

Peirce CS. Letters to Lady Welby. In: Wiener PP, editor. Charles Peirce: selected writings. New York: Dover; 1966.

Peirce CS. In: Houser N, Kloesel C, editors. The essential Peirce. 1st ed. Bloomington: Indiana University Press; 1992.

Petrilli S. Crossing out boundaries with global communication: the problem of the subject. Subj Matter. 2005;2(2):33–48.

Petrilli S. Sign studies and semioethics: communication, translation and values. Berlin: de Gruyter Mouton; 2014.

Petrilli S, Ponzio A. Signs of research on signs. Special issue of Semiotische Berichte. 1998; 22 (3/4).

Petrilli S, Ponzio A. Semiotics unbounded: interpretive routes through the open network of signs. Toronto: University of Toronto Press; 2005.

Pfeffer J, Fong CT. The end of business schools? Less success than meets the eye. Acad Manag Learn Educ. 2002;1(1):78–95.

Pierson FC. The education of American businessmen. New York: McGraw-Hill; 1959.

Pinker S. The blank slate: the modern denial of human nature. Harmondsworth: Penguin; 2003.

Poinsot J. In: Deely J, editor. Tractatus de signis: the semiotic of John Poinsot. 2nd ed. South Bend: St. Augustine's Press; 2013.

Ponzio A. Signs, dialogue and ideology. Trans. Petrilli S. Amsterdam: John Benjamins; 1993.

Ponzio A. The I questioned: Emmanuel Levinas and the critique of occidental reason. Subj Matter. 2006a;3(1):1–45

Ponzio A. The dialogic nature of sign. Trans. Petrilli S. New York: Legas; 2006b.

Popper B. Build-a-bomber: why do so many terrorists have engineering degrees? Slate 29 Dec 2009 http://www.slate.com/articles/health_and_science/science/2009/12/buildabomber.html. Last accessed 17 Nov 2014.

Porter LW, McKibbin LE. Management education and development: drift or thrust into the 21st century? New York: McGraw-Hill; 1988.

Pugès L. European business schools and globalization. In: Morsing M, Rovira AS, editors. Business schoolsand their contribution to society. Los Angeles/London: Sage; 2011. p. 57–62.

Rattasepp S, Bennett T, editors. Gatherings in biosemiotics. Tartu: Tartu University Press; 2012.

Richards IA. The philosophy of rhetoric. Oxford: Oxford University Press; 1937.

Ritchie TD, Sedikides C, Skowronski JJ. Emotions experienced at event recall and the self: implications for the regulation of self-esteem, self-continuity and meaningfulness. Memory. 2015;24(5):577–91.

Romanini V, Fernandex E, editors. Peirce and biosemiotics: a guess at the riddle of life. Dordrecht: Springer; 2014.

Rorty RM, editor. The linguistic turn: essays in philosophical method with two retrospective essays. Chicago: University of Chicago Press; 1967.

Ruesch J, Kees W. Nonverbal communication: notes on the visual perception of human relations. Berkeley: University of California Press; 1956.

Schilhab T, et al., editors. The symbolic species evolved. Dordrecht: Springer; 2012.

Schrag CO. Communicative praxis and the space of subjectivity. West Lafayette: Purdue University Press; 2003.

Sebeok TA. Communication among social bees; porpoises and sonar; man and dolphin. Language. 1963;39:448–66.

Sebeok TA. Perspectives in zoosemiotics. The Hague: Mouton; 1972.

Sebeok TA. Ecumenicalism in semiotics. In: Sebeok, editor. A perfusion of signs. Bloomington/London: Indiana University Press; 1977.

Sebeok TA. The semiotic self. In: The sign and its masters, Austin: University of Texas Press; 1979a.

Sebeok TA. Prefigurements of art. Semiotica. 1979b;27:3–73.

Sebeok TA. Looking in the destination for what should have been sought in the source. In: The sign and its masters. Austin: University of Texas Press; 1979c.

Sebeok TA. I think I am a verb: more contributions to the doctrine of signs. New York: Plenum Press; 1986a.

Sebeok TA. The problem of the origin of language in an evolutionary frame. Lang Sci. 1986b;8(2):168–74.

Sebeok TA. In what sense is language a 'primary modeling system'? In: Broms H, Kaufmann R, editors. Semiotics of culture: proceedings of the 25th symposium of the Tartu-Moscow School of Semiotics, Imatra, Finland, 27th–29th July, 1987. Helsinki: Arator; 1988. p. 67–80.

Sebeok TA. The doctrine of signs. In A sign is just a sign. Bloomington: Indiana University Press; 1991a.

Sebeok TA. Semiotics in the USA. Bloomington: Indiana University Press; 1991b.

Sebeok TA. A sign is just a sign. Bloomington: Indiana University Press; 1991c.

Sebeok TA, editor. Biosemiotica I, special issue of Semiotica. 1999;127(1–4):1–131.

Sebeok TA. Signs, bridges, origins. In: Perron P et al., editors. Semiotics as a bridge between the humanities and the sciences. Toronto: Legas; 2000.

Sebeok TA. Nonverbal communication. In: Cobley P, editor. The Routledge companion to semiotics and linguistics. London: Routledge; 2001a.

Sebeok TA. Global semiotics. Bloomington: Indiana University Press; 2001b.

Sebeok TA. Signs: an introduction to semiotics. Toronto: University of Toronto Press; 2001c.

Sebeok TA. Summing up: in lieu of an introduction. In: Cobley P et al., editors. "Semiotics Continues to Astonish": Thomas A. Sebeok and the doctrine of signs. Berlin: de Gruyter Mouton; 2011. p. 453–9.

Sebeok TA, Cobley P. IASS. In: Cobley P, editor. The Routledge companion to semiotics and linguistics. London: Routledge; 2010.

Sebeok TA, Danesi M. The forms of meaning: modeling systems theory and semiotic analysis. Berlin: Mouton de Gruyter; 2000.

Sebeok TA, Rosenthal R, editors. The Clever Hans phenomenon: communication with horses, whales, apes, and people. New York: New York Academy of Sciences; 1981.

Sebeok TA, Umiker-Sebeok J. "You Know My Method": a juxtaposition of Charles S. Peirce and Sherlock Holmes. Bloomington: Gaslight Publications; 1980.

Sebeok TA, Umiker-Sebeok J, editors. Biosemiotics: the semiotic web 1991. Berlin: Mouton de Gruyter; 1992.

Sedikides C, Green JD. What I don't recall can't hurt me: information negativity versus information inconsistency as determinants of memorial self-defense. Soc Cogn. 2004;22(1):4–29.

Seyfarth RM, Cheney DL. Meaning, reference and intentionality in the natural vocalizations of monkeys. In: Roitblat HL et al., editors. Language and communication: comparative perspectives. Hillsdale: Lawrence Erlbaum Associates; 1993.

Siedentop L. Inventing the individual: the origins of western liberalism. Harmondsworth: Penguin; 2015.

Singh S. The code book: the science of secrecy from ancient Egypt to quantum cryptography. New York: Random; 1999.

Snow CP. The two cultures: the Rede lecture 1959. Cambridge: Cambridge University Press; 1959.

Solomon RC. Business and the humanities: an Aristotelian approach to business ethics. In: Donaldson TJ, Freeman ER, editors. Business as a humanity. New York: Oxford University Press; 1994. p. 45–75.

Sonesson G. Pictorial concepts: inquiries into the semiotic heritage and its relevance for the analysis of the visual world. Lund: Lund University Press/Chartwell-Bratt; 1989.

Starkey K, Tempest S. The business school in ruins? In: Gagliardi P, Czarniawska B, editors. Management education and the humanities. Cheltenham: Edward Elgar; 2006. p. 101–12.

Starkey K, Tiratsoo N. The business school and the bottom line. Cambridge: Cambridge University Press; 2007.

Stashower D. The teller of tales: the life of Arthur Conan Doyle. Harmondworth: Penguin; 2000.

Steiner G. After Babel: aspects of language and translation. London: Oxford University Press; 1975.

Stenmark M. Environmental ethics and policy-making. Aldershot: Ashgate; 2002.

Stern D. Diary of a baby. New York: Basic Books; 1998.

Stjernfelt F. Diagrammatology. An investigation on the borderlines of phenomenology, ontology, and semiotics. Dordrecht: Springer; 2007.

Taylor P. Respect for nature: a theory of environmental ethics. Princeton: Princeton University Press; 1986.

Taylor C. Sources of the self: the making of modernity. Cambridge: Cambridge University Press; 1992.

Tomasello M, Carpenter M, Call J, Behne T, Moll H. Understanding and sharing intentions: the origins of cultural cognition. Behav Brain Sci. 2005;28:675–91.

Tønnessen M. Umwelt ethics. Sign Syst Stud. 2003;31(1):281–99.

Tønnessen M. The biosemiotic glossary project: agent, agency. Biosemiotics. 2015;8(1):125–43.
Trotsky L. Art and revolution: writings on literature, politics, and culture. New York/London: Pathfinder; 1992.
Tuchman G. Wannabe U: inside the corporate university. Chicago/London: Chicago University Press; 2009.
Vološinov VN. Discourse in life and discourse in art. In: Freudianism. Bloomington: Indiana University Press; 1990.
von Foerster H. Through the eyes of the other. In: Steier F, editor. Research and reflexivity. London: Sage; 1991.
von Uexküll J. Theoretische biologie. 2nd ed. Frankfurt: Suhrkamp; 1976.
von Uexküll J. A stroll through the worlds of animals and men: a picture book of invisible worlds. Semiotica. 1992;89(4):319–91.
von Uexküll J. An introduction to umwelt. Semiotica. 2001a;134(1/4):107–10.
von Uexküll J. The new concept of umwelt: a link between science and the humanities. Semiotica. 2001b;134(1/4):111–23.
von Uexküll J. A foray into the worlds of animals and humans. Minneapolis: University of Minnesota Press; 2010.
Von Foerster H. On constructing a reality. In: Understanding understanding, Dordrecht: Springer; 2003.
Wacewicz S, Żywiczyński P. Language evolution: why Hockett's design features are a non-starter. Biosemiotics. 2015;8:29–46.
Watt G. Hard cases, hard times and the humanity of law. In: Bate J, editor. The public value of the humanities. London: Bloomsbury; 2011. p. 197–207.
Weber A. Poetic objectivity: toward an ethics of aliveness. Zeitschrift fur Semiotik. 2016; (forthcoming)
Weber BH, Depew DJ, editors. Evolution and learning: the Baldwin effect reconsidered. Cambridge, MA: MIT Press; 2003.
Welby V. Mother-Sense (1904–1910): a selection. In: Petrilli S, editor. Signifying and understanding: reading the works of Victoria Welby and the Signific movement. Berlin: de Gruyter Mouton; 2009a. p. 670–714.
Welby V. Primal-Sense (1904–1910): a selection. In: Petrilli S, editor. Signifying and understanding: reading the works of Victoria Welby and the Signific movement. Berlin: de Gruyter Mouton; 2009b. p. 715–22.
Welsh D. Codes and cryptography. Oxford: Clarendon Press; 1988.
Wheeler JA (with Ford, Kenneth). Geons, black holes and quantum foam: a life in physics. New York: Norton; 1998.
Wood DJ, Bruner JS, Ross G. The role of tutoring in problem solving. J Child Psychiatry Psychol. 1976;17(2):89–100.
Zeman JJ. Peirce's theory of signs. In: Sebeok TA, editor. A perfusion of signs. Bloomington/ London: Indiana University Press; 1973.

MIX
Papier aus verantwortungsvollen Quellen
Paper from responsible sources
FSC® C105338

If you have any concerns about our products,
you can contact us on
ProductSafety@springernature.com

In case Publisher is established outside the EU,
the EU authorized representative is:
Springer Nature Customer Service Center GmbH
Europaplatz 3, 69115 Heidelberg, Germany

Printed by Libri Plureos GmbH
in Hamburg, Germany